JAMESTOWN'S

Number Power

Robert Mitchell

JAMESTOWN PUBLISHERS

a division of NTC/CONTEMPORARY PUBLISHING GROUP
Lincolnwood, Illinois USA

ISBN: 0-8092-9969-0

Published by Jamestown Publishers,
a division of NTC/Contemporary Publishing Group, Inc.,
4255 West Touhy Avenue,
Lincolnwood (Chicago), Illinois 60712-1975 U.S.A.
1 2 3 4 5 6 7 8 9 VH 16 15 14 13 12 11 10 9 8 7 6 5 4 3 2 1

Table of Contents

Numbers Less than 1

Fractions

To the Student

Welcome to *Number Power Review*.

This book is designed to strengthen your basic math and problem-solving skills by reviewing and mastering topics from whole numbers to algebra. You will develop your number sense working with whole numbers, fractions and decimals, percents, and algebra and geometry. As you work through the lessons of *Number Power Review* you will use estimation and a calculator, you will explore problem-solving methods, and you will graph, analyze, and use data. You will develop math skills that will help you understand and solve all kinds of word problems, especially those related to the workplace, the home, and the classroom. You will feel better prepared for various basic skills tests, including educational tests and employment tests.

Special features of *Number Power Review* include:

- a pretest to help you identify those areas of basic math which you still need to work on

- nine chapters covering core math skills in Building Number Power, and a tenth chapter that uses or extends the skills in Using Number Power

- instruction in using a calculator, estimating answers, and solving problems with mental math shortcuts. The following icons will alert you to problems where using these skills will be especially helpful:

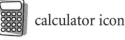 calculator icon

(?) estimation icon

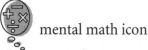 mental math icon

- a review at the end of each chapter to allow you to apply your skills in a test-like format

- two posttests (one with all multiple-choice questions) to help identify those skills you may wish to review and to check your readiness to take other basic skills math tests

To get the most from your work, do each problem carefully. Inside the back cover is a chart to help you keep track of your score on each exercise.

We hope you enjoy *Number Power Review!*

Number Power Review Pretest

This pretest will tell you which chapters of *Number Power Review* you need to work on and which you have already mastered. Do all the problems that you can. There is no time limit. Then use the chart at the end of the test to find the pages where you need work.

1. Write the number 505,005 in words.

Write the inequality that is graphed on each number line.

2.
```
  -5 -4 -3 -2 -1  0  1  2  3  4  5
```
x

3.
```
  -5 -4 -3 -2 -1  0  1  2  3  4  5
```
n

Estimate an answer to each problem.

4.
$$
\begin{array}{r}
793 \\
517 \\
+\ 385 \\
\end{array}
$$

5. $215 \times 89 =$

6. $19\overline{)417}$

Solve each problem.

7.
$$
\begin{array}{r}
2{,}800 \\
-\ \ 946 \\
\end{array}
$$

8.
$$
\begin{array}{r}
173 \\
\times\ \ 26 \\
\end{array}
$$

9. $16\overline{)4{,}208}$

10. Find the value of the expression: $3(\$2.50 - \$1.00) + (\$10.00 \div 4)$.

11. What is the mean price of tickets listed in the table at the right?

Ticket Prices

Section	Price
A	$25
B	$17
C	$12

12. A calculator displays the following number. How is this number read in words?

$$0.025$$

13. Write the following numbers in order, from least to greatest.

$\frac{3}{4}$ 0.34   30% $\frac{2}{3}$ 4%   0.4

14. What percent of the pizza has been eaten?

Write <, >, or = to compare each pair of numbers.

15. $\frac{3}{8}$ _____ $\frac{21}{48}$

16. $4\frac{2}{3}$ _____ $\frac{14}{3}$

17. $2\frac{4}{5}$ _____ $2\frac{11}{15}$

Solve each problem.

18. $2\frac{3}{4}$
   $+ 1\frac{1}{3}$

19. $\frac{2}{3} \times \frac{3}{5} =$

20. $\frac{1}{4} \div \frac{5}{8} =$

21. Keisha bought sixteen cans of soda for a birthday party. Six of these are cans of cola.

a. What fraction of the soda is cola?

b. What is the ratio of cola to flavors that are *not* cola?

Write <, >, or = to compare the pair of decimals.

22. 0.65 _____ 0.605

23. 0.329 _____ 0.48

24. 1.14 _____ 1.042

25. Which expression can be used to find the width of each board shown at right?

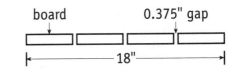

a. $\frac{18 + (3 \times 0.375)}{4}$

d. $\frac{18 - (3 \times 0.375)}{4}$

b. $\frac{18 + (3 \times 0.375)}{3}$

e. $\frac{18 - (4 \times 0.375)}{4}$

c. $\frac{18 - (3 \times 0.375)}{3}$

Solve each problem.

26. 5.08
− 2.92

27. $2.05 \times 0.06 =$

28. $1.4\overline{)2.842}$

29. The Clothes Tree is offering a 25% discount on all women's sweaters. Including a 6% sales tax, what will Selena pay for a sweater that normally sells for $48?

For problems 30 and 31, refer to the circle graph.

30. What percent of the students at Highland Middle School are in the 8th grade?

31. Suppose that next year the enrollment at Highland Middle School increases by 20%. How many students will be enrolled at Highland next year?

HIGHLAND MIDDLE SCHOOL
(enrollment by grade)

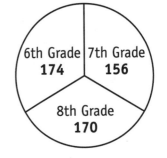

32. a. What is the ratio of the amount that Jonathon saved in May to the amount he saved in January?

b. If Jonathon's take-home pay is $1,500 per month, how much did he save during the 6 months shown on the graph?

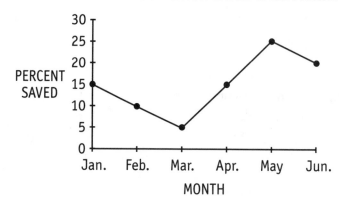

PERCENT OF INCOME JONATHON SAVED EACH MONTH

For problems 33 and 34, refer to the bag of 10 marbles.

33. If Tracy randomly takes one marble from the bag, what is the probability she will take the red marble?

34. After Traci takes one marble and looks at it, she returns it to the bag. Now Brian randomly takes two marbles from the bag. What is the probability that Brian takes two blue marbles?

Solve each equation.

35. $y - 15 = 12$

36. $7x = 63$

37. $\frac{n}{5} = 10$

38. $3x + 5 = 17$

39. $5y + 2y = 21$

40. $\frac{3}{7} = \frac{w}{28}$

41. a. Graph the equation $y = 2x - 2$. Start by plotting the three points that have x values −2, 1, and 4.

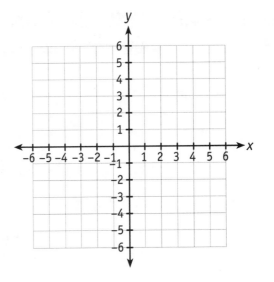

    b. What are the intercepts of your graphed line?

      x-intercept = (,)

      y-intercept = (,)

    c. What is the slope of your graphed line?

42. Blair Street intersects 4th Avenue at an acute angle of 43°. What is the measure of the obtuse angle at which Blair intersects 4th?

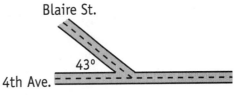

Blaire St.

43°

4th Ave.

43. Hal is standing 40 feet away from a 36 foot tree. If the distance from the top of the tree to the top of Hal's head is 50 feet, how tall is Hal? (**Hint:** Use the Pythagorean theorem.)

50 ft   36 ft

40 ft

44. The cylindrical oil drum shown at the right is half full of oil.

    a. To the nearest cubic foot, what is the volume of the oil drum?
(Use $\pi \approx 3.14$ or $\pi \approx \frac{22}{7}$.)

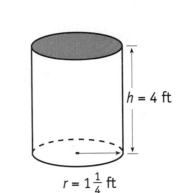

$h = 4$ ft

$r = 1\frac{1}{4}$ ft

    b. To the nearest gallon, how much oil is in the drum now?
(1 cu ft $\approx 7\frac{1}{2}$ gal)

Number Power Review Pretest Chart

This pretest can be used to identify math skills in which you are already proficient. If you miss only one question in a chapter, you may not need further study in that chapter. However, before you skip any chapters, we recommend that you complete the review test at the end of that chapter. For example, if you miss one question about fractions, do the Fractions Review (pages 110–111) before beginning Decimals. This longer inventory will be a more precise indicator of your skill level.

PROBLEM NUMBERS	SKILL AREA	PRACTICE PAGES
1	writing larger numbers	9
2, 3	graphing inequalities	26–27
4, 5, 6	estimating with whole numbers	12–19
7, 8, 9	calculating with whole numbers	32–35, 40–43
10	evaluating an expression	46–49
11	finding a mean	50–51
12	reading a decimal number	56–57
13	comparing numbers less than 1	72–73
14	understanding percent	70–71
15, 16, 17	comparing fractions, mixed numbers	82
18, 19, 20	calculating with fractions, mixed numbers	83–85, 94–99
21	relating fractions and ratios	104–105
22, 23, 24	comparing decimals	58
25	understanding set-up questions	36–37, 120–123
26, 27, 28	calculating with decimals	116–123
29	finding a percent of a whole	134–135, 142–143
30, 31	working with a circle graph	74–75
32	working with a line graph	128–129
33	finding the probability of one event	166–167
34	finding the probability of two events	170–171
35, 36, 37	solving one-step equations	176–179
38, 39, 40	solving multistep equations	180–181
41	graphing a linear equation	196–199
42	working with angles	204–209
43	using the Pythagorean theorem	216–217
44	finding volume	224–225

Building
Number
Power

NUMBER SENSE

Writing Familiar Numbers

Most numbers that we use in daily life have four or fewer digits. The number 1,326 is an example of a four-digit number.

Notice how the **value** of each digit is determined by its **place** in the number.

The First Four Place Values

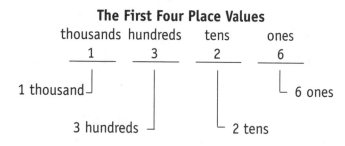

thousands hundreds   tens   ones
   1   3   2   6

1 thousand⌟

3 hundreds ⌟

2 tens

6 ones

Math Tip
A number is the sum of its parts.
1,326

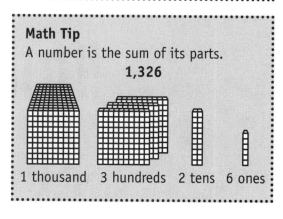

1 thousand   3 hundreds   2 tens   6 ones

Zero is used as a **placeholder** to show that there are no ones, tens, or hundreds.

No Ones	No Tens	No Hundreds
3, 6 4 0	1, 5 0 9	2, 0 7 5

3 thousands⌟   ⌞**0 ones**   1 thousand⌟   ⌞9 ones   2 thousands⌟   ⌞5 ones

6 hundreds ——⌞ ⌞4 tens   5 hundreds —— ⌞**0 tens**   **0 hundreds** —— ⌞ ⌞7 tens

Look at this example and the rules that follow.
The number 1,973 is read "one thousand, nine hundred seventy-three."

- Do not use the word *and* when writing or saying whole numbers.
- Write a hyphen (-) in compound words such as twenty-one (21) and ninety-nine (99).
- Place a comma after the word *thousand,* but not after the word *hundred.*

Write each number in words.

1. 46 _____

2. 381 _____

3. 1,508 _____

4. 7,062 _____

Writing Larger Numbers

In larger numbers, commas separate digits into groups of three. Notice each pattern of hundreds, tens, and ones in the 3-digit groups.

Millions	Thousands	Ones

____ ____ ____, ____ ____ ____, ____ ____ ____

100s 10s 1s   100s 10s 1s   100s 10s 1s

To read a larger number, read each group of digits separately.

239,710,579 is read 239 *million*, 710 *thousand*, 579.

millions｜ ones
　　　thousands

(**Note:** Commas are used to separate millions from thousands and thousands from ones.)

Math Tip

To strengthen skills for reading larger numbers, memorize these:

Number	Read as
1,000	1 thousand
10,000	10 thousand
100,000	100 thousand
1,000,000	1 million
10,000,000	10 million
100,000,000	100 million

Match each number with its word meaning.

Number

_____ **1.** 306,000

_____ **2.** 3,600,000

_____ **3.** 36,000,000

_____ **4.** 3,006,000

_____ **5.** 360,000

_____ **6.** 30,060,000

Word Meaning

a. three million, six thousand

b. thirty-six million

c. three million, six hundred thousand

d. thirty million, sixty thousand

e. three hundred six thousand

f. three hundred sixty thousand

Write each number using digits.

7.　sixty-one thousand, eight hundred seventy-two _____

8.　four hundred sixty-two thousand, eight hundred _____

9.　five million, eight hundred thirty-five thousand _____

Write each number in words.

10.　diameter of the earth: 7,926 miles _____

11.　distance of the moon from the earth: 237,300 miles _____

12.　distance of the sun from the earth: 92,900,000 miles _____

Comparing and Ordering Numbers

Symbols are used to show the relationship between two numbers.

Symbol	Meaning	Example	Read as
>	is greater than	9 > 5	"9 is greater than 5"
<	is less than	3 < 11	"3 is less than 11"
=	is equal to	6 = 6	"6 is equal to 6"
≠	is not equal to	8 ≠ 12	"8 is not equal to 12"

To compare two numbers, compare digits from left to right. Be sure to compare digits that have the same place value.

EXAMPLE 1

Compare 126 and 98.

$\boxed{\begin{matrix}1\end{matrix}}$ 2 6   Because 1 > 0,

9 8   126 > 98.

EXAMPLE 2

Compare 198 and 249.

$\boxed{\begin{matrix}1\\2\end{matrix}}$ 9 8   Because 1 < 2,

4 9   198 < 249.

EXAMPLE 3

Compare 361 and 347.

3 $\boxed{\begin{matrix}6\\4\end{matrix}}$ 1   From the left: 3 = 3,

3   7   6 > 4, so 361 > 347.

Write >, <, or = to compare each pair of numbers below.

1. 38 _____ 42   30 _____ 30 $129 _____ $124
2. 819 _____ 822   1,035 _____ 872 $2,487 _____ $2,574
3. 7,975 _____ 7,975 $5,052 _____ $5,061   25,435 _____ 25,431
4. $6.38 _____ $6.78 $14.61 _____ $13.73 $25.18 _____ $34.06

Look at the employee loan list. Rewrite the list in order of loan amount. List the largest amount first (1st) and so on.

Loan Amount

5. 1st: _____

7. 3rd: _____

9. 5th: _____

Loan Amount

6. 2nd: _____

8. 4th: _____

10. 6th: _____

Employee Name	Loan Amount
Cheung	$4,867
Cooke	$5,932
Ferguson	$6,779
Herzberg	$5,589
Nason	$6,780
Ramirez	$4,814

Understanding Number Sentences

A **number sentence** is a mathematical statement that compares numbers of **numerical expressions**—numbers joined by +, −, ×, and ÷ signs. A number sentence may be an **equation** or an **inequality**.

Equation	Inequalities
$24 = 14 + 10$	$106 > 80 + 9$ $63 < 90 - 20$ $14 \neq 7 + 9$

- An **equation** contains the **equals sign** (=).

- An **inequality** contains an **inequality sign** (>, <, or ≠).

To see if a number sentence is true, perform all operations (+, −, ×, or ÷) and then compare the resulting two numbers.

Which of the following number sentences are true?

a.	$23 + 10 > 35$	**False,** because $23 + 10 = 33$, and $33 < 35$.
b.	$34 - 12 = 20$	**False,** because $34 - 12 = 22$, *not* 20.
c.	$30 \div 6 < 6$	**True,** because $30 \div 6 = 5$, and $5 < 6$.
d.	$14 \times 3 \neq 51$	**True,** because $14 \times 3 = 42$, and $42 \neq 51$.

For each number sentence below, perform the indicated operation. Then circle *T* if the sentence is true or *F* if it is false.

1. $9 < 5 + 7$	*T* or *F*	$14 - 9 > 3$	*T* or *F*	$13 + 12 \neq 25$	*T* or *F*
2. $11 > 17 - 6$	*T* or *F*	$15 - 7 = 9$	*T* or *F*	$3 \times 4 > 14$	*T* or *F*
3. $20 \neq 2 \times 9$	*T* or *F*	$9 < 56 \div 7$	*T* or *F*	$42 \div 6 > 7$	*T* or *F*
4. $5 \times 8 > 38$	*T* or *F*	$36 \div 4 < 8$	*T* or *F*	$65 > 9 \times 7$	*T* or *F*

Write a *whole number* to make each equation or inequality true. For each inequality, more than one answer is correct.

5. $5 + \underline{\hspace{1cm}} = 12$	$7 \times \underline{\hspace{1cm}} = 42$	$9 + 8 \neq 12 + \underline{\hspace{1cm}}$
6. $45 \div 9 > \underline{\hspace{1cm}}$	$30 - \underline{\hspace{1cm}} < 5 \times 2$	$4 \times 8 < \underline{\hspace{1cm}}$
7. $14 - 8 > \underline{\hspace{1cm}}$	$3 \times 4 > \underline{\hspace{1cm}}$	$25 \div \underline{\hspace{1cm}} > 5$
8. $9 \times \underline{\hspace{1cm}} = 54$	$8 - \underline{\hspace{1cm}} > 3$	$32 \div \underline{\hspace{1cm}} = 4$

Rounding Whole Numbers

Often, it is useful to **estimate**—to give a number that is "about equal" to an exact amount. One way to estimate is to use round numbers.

A newspaper says that 800 people attended a picnic. Actually, 829 people were there. The number 800 is called a **round number.** To the hundreds place, 829 **rounds to** 800. To the thousands place, 829 rounds to 1,000.

> **Math Tip**
> A round number usually ends in one or more zeros.

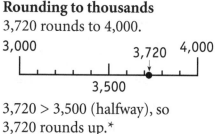

Rounding to tens
25 rounds to 30.

25 is halfway between 20 and 30, so 25 rounds up.*

Rounding to hundreds
138 rounds to 100.

138 < 150 (halfway), so 138 rounds down.**

Rounding to thousands
3,720 rounds to 4,000.

3,720 > 3,500 (halfway), so 3,720 rounds up.*

*A number that's *halfway or more* to the larger round number rounds *up* to the larger number.

**A number that's *less than halfway* to the larger round number rounds *down* to the smaller number.

....

Round each number. Circle each answer choice.

To the nearest ten	To the nearest hundred	To the nearest thousand
1. 42: 40 or 50	175: 100 or 200	1,198: 1,000 or 2,000
2. 67: 60 or 70	750: 700 or 800	4,723: 4,000 or 5,000

Round each number to the nearest ten.

3. 35 _____   73 _____   19 _____   84 _____   95 _____

Round each number to the nearest hundred.

4. 109 _____   350 _____   284 _____   420 _____   834 _____

Round each number to the nearest thousand.

5. 3,058 _____   1,942 _____   7,498 _____   4,500 _____   6,056 _____

Rounding Dollars and Cents

You often round money to the nearest $1.00 or $10.00.

Rounding to the Nearest Dollar
Look at the number of cents.

$0.<u>83</u> rounds to $1.00
↑

Round up if
greater than or
equal to 50¢.

$3.<u>29</u> rounds to $3.00
↑

Round down if
less than 50¢.

$12.<u>50</u> rounds to $13.00
↑

Round up if
greater than or
equal to 50¢.

Rounding to the Nearest Ten Dollars
Look at the number of dollars.

$<u>9</u>.18 rounds to $10.00
↑

Round up if greater than
or equal to $5.

$<u>4</u>1.84 rounds to $40.00
↑

Round down if less than $5.

$17<u>5</u>.00 rounds to $180.00
↑

Round up if greater than or
equal to $5.

Round each purchase to the nearest dollar.

1.

$ 1.89

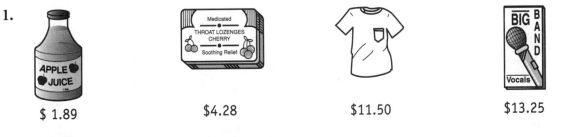

$4.28 $11.50 $13.25

Round each purchase to the nearest ten dollars.

2.

$19.75 $32.27

$107.50

$88.88

Round these amounts.

3. $529 to the nearest hundred dollars

4. $1,675 to the nearest hundred dollars

5. $3,819 to the nearest thousand dollars

6. $14,395 to the nearest thousand dollars

Estimating Sums

Estimating can be your most important math tool. You estimate to

- find a "close" answer when that's all you need
- check calculation accuracy
- check calculator answers
- help you choose answers on tests

To estimate a sum, round each number. Then add the rounded numbers.

EXAMPLE 1 Estimate: 119 + 72

Round each number to the tens place. Then add.

$$
\begin{array}{r}
119 \rightarrow 120 \\
+\ 72 \rightarrow +\ 70 \\
\hline
190
\end{array}
$$

EXAMPLE 2 Estimate: 479 + 318

Round each number to the hundreds place. Then add.

$$
\begin{array}{r}
479 \rightarrow 500 \\
+318 \rightarrow +300 \\
\hline
800
\end{array}
$$

EXAMPLE 3 Estimate: 1,191 + 594

Round each number to the hundreds place. Then add.

$$
\begin{array}{r}
1,191 \rightarrow 1,200 \\
+\ 594 \rightarrow +\ 600 \\
\hline
1,800
\end{array}
$$

Estimate these sums. Any reasonable estimate is acceptable.

1. $\begin{array}{r} 56 \\ +29 \end{array}$ $\begin{array}{r} 73 \\ +52 \end{array}$ $\begin{array}{r} 327 \\ +188 \end{array}$ $\begin{array}{r} 493 \\ +218 \end{array}$

2. $\begin{array}{r} 1,193 \\ +\ 720 \end{array}$ $\begin{array}{r} 3,265 \\ +\ 514 \end{array}$ $\begin{array}{r} 3,274 \\ +1,865 \end{array}$ $\begin{array}{r} 5,878 \\ +2,387 \end{array}$

3. $\begin{array}{r} \$7.89 \\ +\ 5.14 \end{array}$ $\begin{array}{r} \$13.45 \\ +\ 4.89 \end{array}$ $\begin{array}{r} \$125.99 \\ +\ 62.84 \end{array}$ $\begin{array}{r} \$119.23 \\ +\ 71.84 \end{array}$

Estimate an answer to each problem. Using only your estimate, choose the exact answer.

4. $\begin{array}{r} 391 \\ 207 \\ +193 \end{array}$
    a. 641
    b. 791
    c. 921

5. $\begin{array}{r} \$19.59 \\ 14.98 \\ +10.25 \end{array}$
    a. $34.92
    b. $40.32
    c. $44.82

6. $\begin{array}{r} 2,129 \\ 1,082 \\ +\ 539 \end{array}$
    a. 3,750
    b. 4,470
    c. 5,290

Using Clustering to Estimate a Large Sum

Multiplication can often be used to estimate the value of a large sum. When a group of numbers **clusters** around a common value, you estimate the total by *multiplying* the common value by the number of values in the group.

EXAMPLE 1 Estimate the sum.

$$\begin{array}{r} 1{,}198 \\ 1{,}047 \\ 983 \\ +\ 856 \\ \hline \end{array}$$

The numbers cluster around 1,000. A good estimate is

Total = 1,000 × 4
= 4,000

(**Note:** When possible, choose a round number ending in zeros as the *common value.*)

EXAMPLE 2 Estimate the total number of houses sold during the months shown.

The numbers cluster around 300. A good estimate is

Total = 300 × 4
= 1,200 houses

Houses Sold	
Month	**Sales**
March	321
April	287
May	296
June	345
Est. Total:	

(?) **For each problem, find the common value that the numbers cluster around. Then estimate a sum.**

1. $\begin{array}{r} 3{,}275 \\ 3{,}092 \\ 2{,}947 \\ +\ 2{,}890 \\ \hline \end{array}$

   a. Common value: _____

   b. Estimated sum: _____

2. Estimate the total hamburger sales.

Day	Sales
Monday	693
Tuesday	746
Wednesday	613
Thursday	694
Friday	715
Est. Total:	

3. $\begin{array}{r} \$2{,}147 \\ 1{,}908 \\ 2{,}009 \\ +\ 1{,}846 \\ \hline \end{array}$

   a. Common value: _____

   b. Estimated value: _____

5. Estimate the total number of miles driven.

Driver	Miles
Kushmir	1,983
Atwood	2,176
Burns	1,857
Rayas	2,209
Est. Total:	

4. $\begin{array}{r} \$489.50 \\ \$473.80 \\ \$536.90 \\ \$518.00 \\ +\ \$507.75 \\ \hline \end{array}$

   a. Common value: _____

   b. Estimated sum: _____

Estimating Differences

To estimate a difference, round each number to the same place. Then subtract the rounded numbers.

> **Math Tip**
> Estimating works well on multiple-choice questions when the answer choices are not too close together in value. (See problems 4–6 on page 14.)
>
> **Estimating helps here:**   **but not here!**
>
> Subtract: Subtract:
>
> | | Estimate | | | Estimate | |
> |-------|----------|-----------|-------|----------|----------|
> | 301 | 300 | a. 81 | 407 | 400 | a. 208 |
> | − 196 | − 200 | *b. 105 | − 189 | − 200 | b. 213 |
> | | 100 | c. 123 | | 200 | *c. 218 |
>
> Estimating works well   The answer choices are too
> because the answer choices   close in value for the
> are not close in value.   estimate to help.
>
> *correct answer

Estimate these differences. Any reasonable estimate is acceptable.

1.
37	172	67	442
− 19	− 99	− 26	− 116

2.
1,314	1,168	3,289	5,110
− 829	− 593	− 1,317	− 2,967

3.
$8.93	$17.18	$113.12	$142.50
− 4.16	− 9.24	− 71.92	− 98.78

Use estimation to help you choose each exact answer below. Circle the number of the problem for which estimating doesn't help.

4.
686	a. 268
− 298	b. 388
	c. 528

5.
$48.64	a. $28.66
− 19.98	b. $29.06
	c. $29.76

6.
3,963	a. 1,294
− 1,979	b. 1,674
	c. 1,984

Estimating Products

One way to estimate a product is to round each number to its highest place; each number will then contain a single digit and one or more zeros. Then multiply the rounded numbers.

Math Tip
To multiply numbers that end in zero

- multiply the nonzero digits
- add on the number of zeros in the two numbers

$$\begin{array}{r} 300 \\ \times\ 20 \\ \hline 6{,}000 \\ \uparrow \\ 3 \times 2 \end{array} \quad \begin{array}{l} 2\ \text{zeros} \\ +\ 1\ \text{zero} \\ \hline \leftarrow \quad 3\ \text{zeros} \end{array}$$

EXAMPLE 1

Estimate: 94×7

Round 94 and multiply.

$$\begin{array}{ccc} 94 & \to & 90 \\ \times\ 7 & \to & \times\ 7 \\ & & \hline 630 \end{array}$$

EXAMPLE 2

Estimate: $\$79 \times 21$

Round each number to the nearest ten. Then multiply.

$$\begin{array}{ccc} \$79 & \to & \$80 \\ \times\ 21 & \to & \times\ 20 \\ & & \hline \$1{,}600 \end{array}$$

EXAMPLE 3

Estimate: 213×187

Round each number to the nearest hundred. Then multiply.

$$\begin{array}{ccc} 213 & \to & 200 \\ \times 187 & \to & \times\ 200 \\ & & \hline 40{,}000 \end{array}$$

Estimate these products. Any reasonable estimate is acceptable.

1.
$$\begin{array}{r} 37 \\ \times\ 8 \\ \hline \end{array} \qquad \begin{array}{r} 81 \\ \times\ 9 \\ \hline \end{array} \qquad \begin{array}{r} 41 \\ \times 32 \\ \hline \end{array} \qquad \begin{array}{r} 58 \\ \times 19 \\ \hline \end{array}$$

2.
$$\begin{array}{r} 112 \\ \times\ 23 \\ \hline \end{array} \qquad \begin{array}{r} 273 \\ \times\ 49 \\ \hline \end{array} \qquad \begin{array}{r} 302 \\ \times 116 \\ \hline \end{array} \qquad \begin{array}{r} 593 \\ \times 287 \\ \hline \end{array}$$

3.
$$\begin{array}{r} \$7.89 \\ \times\ 8 \\ \hline \end{array} \qquad \begin{array}{r} \$9.31 \\ \times\ 6 \\ \hline \end{array} \qquad \begin{array}{r} \$56.99 \\ \times\ 11 \\ \hline \end{array} \qquad \begin{array}{r} \$89.75 \\ \times\ 19 \\ \hline \end{array}$$

For each problem below, choose the best estimate of the product.

4.
$$\begin{array}{r} 79 \\ \times 61 \\ \hline \end{array}$$
   a. 100×50
   b. 80×60
   c. 80×50

5.
$$\begin{array}{r} 113 \\ \times\ 98 \\ \hline \end{array}$$
   a. 125×100
   b. 100×100
   c. 110×100

6.
$$\begin{array}{r} \$139.87 \\ \times\ 49 \\ \hline \end{array}$$
   a. $\$140 \times 50$
   b. $\$150 \times 50$
   c. $\$100 \times 50$

Estimating Quotients

Dividing by a One-Digit Number

To estimate when dividing by a one-digit number, use **compatible numbers.** Compatible numbers are numbers that divide evenly, with no remainder. Compatible numbers come from the basic multiplication facts.

$$4\overline{)28}^{\,7}$$ 4 and 28 are compatible numbers.

To estimate a quotient when you are dividing by a one-digit divisor

- round the first digit or two of the dividend to a number that is compatible with the divisor
- write zeros for the other digits of the dividend
- divide the compatible digits and write zeros as placeholders in the quotient

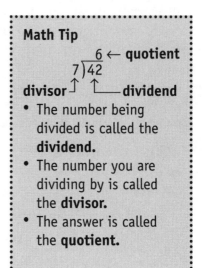

Math Tip

$$7\overline{)42}^{\,6} \leftarrow \textbf{quotient}$$

divisor ↰ ↱ dividend

- The number being divided is called the **dividend.**
- The number you are dividing by is called the **divisor.**
- The answer is called the **quotient.**

EXAMPLE 1

Problem: $7\overline{)37}$

Recall: $7 \times 5 = 35$

Round 37 to 35.

Divide: $7\overline{)35}^{\,5}$

Estimate: 5

EXAMPLE 2

Problem: $8\overline{)719}$

Recall: $8 \times 9 = 72$

Round 719 to 720.

Divide. Then write a placeholding zero.

$$8\overline{)720}^{\,90}$$

Estimate: 90

EXAMPLE 3

Problem: $6\overline{)5,146}$

Recall: $6 \times 8 = 48$

Round 5,146 to 4,800.

Divide. Then write placeholding zeros.

$$6\overline{)4,800}^{\,800}$$

Estimate: 800

(**Note:** You can also round 5,146 to 5,400. That would give an estimate of 900.)

Estimate these quotients. Any reasonable estimate is acceptable.

1. $6\overline{)43}$ $5\overline{)29}$ $7\overline{)24}$ $9\overline{)\$38}$

2. $7\overline{)489}$ $8\overline{)\$251}$ $3\overline{)817}$ $4\overline{)\$315}$

3. $2\overline{)\$3,994}$ $5\overline{)1,439}$ $7\overline{)\$5,179}$ $8\overline{)7,357}$

Dividing by a Two-Digit Number

One way to estimate when the divisor contains more than one digit is to round both the divisor and the dividend.

- Round to numbers that make the division easy for you. Usually, this means rounding to numbers where the first digits of the divisor and dividend are compatible numbers.
- Divide and write zeros as placeholders when needed.

EXAMPLE 1

Problem: $23\overline{)827}$

Think: $8 \div 2 = 4$

Round 23 to 20 and 827 to 800.

Divide: $20\overline{)80\emptyset}$ with 40 above

STEP 1 Cross out one zero in both the divisor and the dividend.

STEP 2 Divide 80 by 2.

Estimate: 40

EXAMPLE 2

Problem: $89\overline{)4,609}$

Think: $45 \div 9 = 5$

Round 89 to 90 and 4,609 to 4,500.

Divide: $9\emptyset\overline{)4,50\emptyset}$ with 50 above

STEP 1 Cross out one zero in both the divisor and the dividend.

STEP 2 Divide 450 by 9.

Estimate: 50

Estimate these quotients.

4. $21\overline{)86}$ $32\overline{)\$90}$ $11\overline{)20}$ $19\overline{)\$80}$

5. $39\overline{)129}$ $51\overline{)488}$ $21\overline{)\$837}$ $17\overline{)527}$

6. $82\overline{)4,390}$ $28\overline{)\$7,148}$ $41\overline{)3,840}$ $72\overline{)\$12,343}$

Data Highlight: Reading a Table

Nutritional Information for Selected Fast-Food Sandwiches (Typical values are given.)			
Food Item	Calories	Protein (g*)	Fat (g*)
Hamburger	428	28	19
Cheeseburger	470	31	23
Fish Sandwich	344	33	10
Chicken Sandwich	378	32	12
Ham Sandwich	451	30	21

*g = grams   A gram is a metric weight unit.
A raisin weighs about one gram.

The nutritional information above is presented in the form of a **table.** A table usually has these features: a **title** that describes the table; horizontal **rows** and vertical **columns;** and **headings** for the rows and columns.

- Rows are read across.   Cheeseburger   470   31   23
- Columns are read from top to bottom.   **Calories**
  428
  470
  344
  378
  451

EXAMPLE 1   How many grams of protein does a chicken sandwich contain?

Find the intersection of the row labeled Chicken Sandwich and the column labeled Protein.

  Protein
  28
  31
  33
Chicken Sandwich 378 32 12
  30

ANSWER: 32 grams

EXAMPLE 2   Which of the listed sandwiches is lowest in fat?

Scan down the column labeled Fat and choose the smallest number: 10. Read the label of the row that contains the 10.

  Fat
  19
  23
Fish Sandwich 344   33 10
  12
  21

ANSWER: Fish Sandwich

For problems 1–5, refer to the nutrition table on page 20.

1. How many calories does a cheeseburger contain?

2. How many grams of fat does a chicken sandwich contain?

3. According to the table, which sandwich has the

   a. most calories? _____   c. most fat? _____
   b. least amount of protein? _____   d. fewest calories? _____

4. A sandwich with a high fat content tends to have a high amount of

   a. protein
   b. weight
   c. calories

5. A sandwich with a high calorie content tends to have a high amount of

   a. fat
   b. protein
   c. vitamins

For problems 6–9, refer to the portion of the federal income tax table shown below.

If Taxable income is—		And you are—			
At least	But less than	Single	Married filing jointly	Married filing separately	Head of a household
		Your Tax is—			
$20,800	$20,850	$3,124	$3,124	$3,225	$3,124
20,850	20,900	3,131	3,131	3,239	3,131
20,900	20,950	3,139	3,139	3,253	3,139
20,950	21,000	3,146	3,146	3,267	3,146
21,000	21,050	3,154	3,154	3,281	3,154

6. Ben is single and has a taxable income of $20,945. How much tax does he owe?

   Tax owed: _____

7. Anne is married and is filing jointly with her husband. If their taxable income is $21,030, how much tax do they owe?

   Tax owed: _____

8. Julia is "head of a household." If she owes $3,131 in taxes, her taxable income is between

   $_____ and $_____.

9. José is married but is filing separately from his wife. If José must pay $3,253 in taxes, his taxable income is between

   $_____ and $_____.

Number Highlight: Negative Numbers

Along with the **positive numbers** you've been studying in this chapter, there are also **zero** and the **negative numbers.**

- Positive numbers are numbers greater than zero.
- Negative numbers are numbers less than zero.

You may already know some of the common uses of negative numbers:

- reading temperatures below zero
- recording yardage losses in football
- reporting drops in stock market prices
- measuring distances below sea level

Always write a negative (minus) sign in front of a negative number.

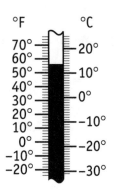

Current Temperature
−12° Celsius

Stock Market Changes
IBM −3

The Number Line

A **number line** looks like a thermometer scale on its side. Negative numbers are written to the left of zero; positive numbers are written to the right of zero.

Negative Numbers   Zero   Positive Numbers

−12 −10 −8 −6 −4 −2   0   2   4   6   8   10   12

Numbers on a number line get larger as you move from left to right. Every negative number is less than 0. Every positive number is greater than 0.

EXAMPLES   −8 is less than −6   6 is greater than −8   −1 is less than 1

Addition and subtraction of negative numbers are easily visualized on a number line. To *add,* start at the first number and move the second number of spaces to the answer.

- If the second number is positive, move to the right of the first number.
- If the second number is negative, move to the left of the first number.

EXAMPLE 1   Add: −3 + 8

Start   Move 8 spaces right.

−6 −4 −3 −2   0   2   4   5   6

ANSWER: 5

EXAMPLE 2   Add: 4 + (−6)

Move 6 spaces left.   Start

−6 −4 −2   0   2   4   6

ANSWER: −2

To *subtract,* count the number of spaces between the two numbers.

- The answer is positive when you subtract a smaller number from a larger number.
- The answer is negative when you subtract a larger number from a smaller number.

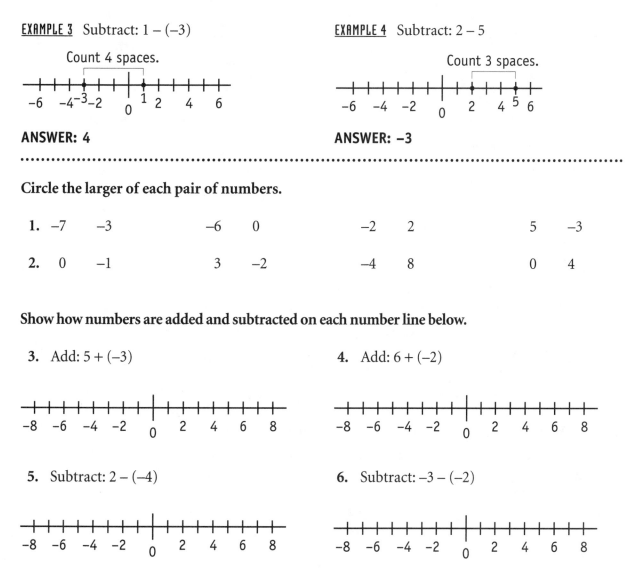

EXAMPLE 3 Subtract: 1 − (−3)

Count 4 spaces.

ANSWER: 4

EXAMPLE 4 Subtract: 2 − 5

Count 3 spaces.

ANSWER: −3

Circle the larger of each pair of numbers.

1. −7   −3   −6   0   −2   2   5   −3

2. 0   −1   3   −2   −4   8   0   4

Show how numbers are added and subtracted on each number line below.

3. Add: 5 + (−3)

4. Add: 6 + (−2)

5. Subtract: 2 − (−4)

6. Subtract: −3 − (−2)

Solve each problem. Sketch number lines if you find it helpful.

7. At 2:00 A.M. Sunday, the temperature in Peoria was −5°F. By 6:00 A.M., the temperature had risen by 8°. What was the temperature in Peoria at 6:00 A.M. Sunday?

8. An airplane flying at 800 feet above sea level detects a submarine directly below, submerged at a depth of −200 feet (200 feet below sea level). How high is the plane above the submarine?

More About Inequalities

An **inequality** is a statement that two numbers are *not* equal.

Four comparison symbols are used with inequalities when a letter is used to stand for more than one possible value.

Symbol	Meaning	Example	Read as
<	is less than	$n < 8$	n is less than 8.
>	is greater than	$m > 10$	m is greater than 10.
≤	is less than *or* equal to	$r \leq 12$	r is less than *or* equal to 12.
≥	is greater than *or* equal to	$s \geq 15$	s is greater than *or* equal to 15.

In each inequality, the value of the letter or **variable** can be *any value* that makes the inequality true.

- $n < 8$ n can be any value less than 8.
  Example values: $n = 5$, $n = 0$ or $n = -7.5$.

- $m > 10$ m can be any value greater than 10.
  Example values: $m = 10.5$, $m = 11$, or $m = 21\frac{1}{4}$.

- $r \leq 12$ r can be any value less than *or* equal to 12.
  Example values: $r = 12$, $r = 9$, or $r = -2.75$.

- $s \geq 15$ s can be any value greater than *or* equal to 15.
  Example values: $s = 15$, $s = 20$, or $s = 30\frac{1}{2}$.

For each inequality, circle any allowed value for the letter or variable. Two are done as examples.

1. **a.** $n < 9$ (−4) (0) (6)   9   10   12   **b.** $p < 11$ −1   0   4   10   12   14

2. **a.** $t > 10$ −3   1   7   10   13   25   **b.** $x > 7$ −2   0   7   9   18   26

3. **a.** $m \leq 15$ (−3) (1) (15)   18   27   39   **b.** $n \leq 5$ −6   1   4   5   7   10
   [m can be equal to 15 *or* m can be any number less than 15.]   [n can be equal to 5 *or* n can be any number less than 5.]

4. **a.** $y \geq 12$ −3   1   12   18   27   39   **b.** $x \geq 8$ −9   4   8   13   18   21
   [y can be equal to 12 *or* y can be any number greater than 12.]   [x can be equal to 8 *or* x can be any number greater than 8.]

Range of Values

Two comparison symbols can be used together to describe a range of values. The lowest end of the range is usually written as the first number of the inequality.

Inequality	Meaning	Example Values
$5 < n < 8$	n is greater than 5 *and* less than 8.	$n = 6$, $n = 7$, $n = 7.5$
$-3 \le m < 2$	m is greater than or equal to -3 *and* less than 2.	$m = -3$, $m = 0$, $m = 1$
$0 < x \le 5$	x is greater than 0 *and* less than or equal to 5.	$x = \frac{3}{4}$, $x = 1$, $x = 4.2$, $x = 5$
$-6 \le y \le 6$	y is greater than or equal to -6 *and* less than or equal to 6.	$y = -6$, $y = 0$, $y = 2.9$, $y = 6$

Write the meaning of each inequality.

5. $-3 < x < 7$ _____

6. $-4 \le y < 20$ _____

7. $6 < p \le 17$ _____

8. $-9 \le x \le 13$ _____

For each inequality, three possible values are given. If the value is correct (that is, if the value makes the inequality a true statement), circle *yes*. If not, circle *no*.

9. $-1 < x < 6$ **a.** $x = 4$ *yes no*   **b.** $x = -1$ *yes no*   **c.** $x = 0$ *yes no*

10. $-3 \le n < 0$ **a.** $n = 0$ *yes no*   **b.** $n = 1$ *yes no*   **c.** $n = -4$ *yes no*

11. $1 < m \le 9$ **a.** $m = -2$ *yes no*   **b.** $m = 0$ *yes no*   **c.** $m = 5$ *yes no*

12. $-7 \le p \le 7$ **a.** $p = -3$ *yes no*   **b.** $p = 0$ *yes no*   **c.** $p = 3$ *yes no*

Showing an Inequality on a Number Line

You can use a number line to show an inequality or range of values.

- A solid line is drawn through all the values in the range.

- An open circle ○ means that the number is *not* included in the range.

- A solid circle ● means that the number *is* included in the range.

EXAMPLE 1 Inequality: $n > -4$

Values of n

Graph:

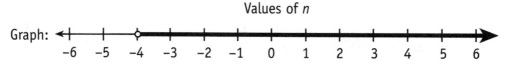

Meaning: n is greater than -4. [n cannot have the value -4.]

EXAMPLE 2 Inequality: $x \geq 0$

Values of x

Graph:

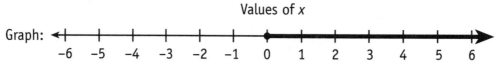

Meaning: x is greater than or equal to 0. [x can have the value 0.]

EXAMPLE 3 Inequality $-5 < s \leq 4$

Values of s

Graph:

Meaning: s is greater than -5 and less than or equal to 4.

EXAMPLE 4 Inequality: $-5 < v < 3$

Values of v

Graph:

Meaning: v is greater than -5 and less than 3.

Circle the letter for the inequality graphed on each number line.

1.

Values of *t*

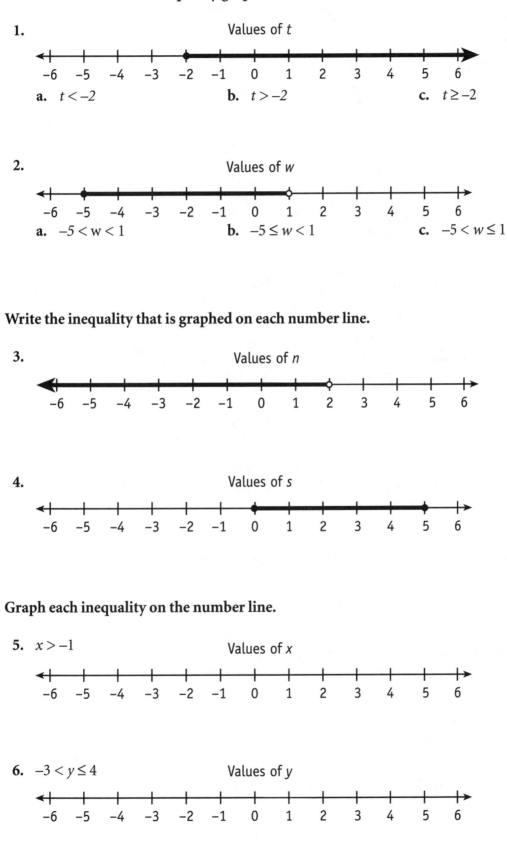

a. $t < -2$ b. $t > -2$ c. $t \geq -2$

2.

Values of *w*

a. $-5 < w < 1$ b. $-5 \leq w < 1$ c. $-5 < w \leq 1$

Write the inequality that is graphed on each number line.

3.

Values of *n*

4.

Values of *s*

Graph each inequality on the number line.

5. $x > -1$

Values of *x*

6. $-3 < y \leq 4$

Values of *y*

Calculator Spotlight

The Calculator Keyboard

Learning to use a calculator quickly and confidently is an important math skill. The calculator below should be similar to one you've seen or one you may be using.

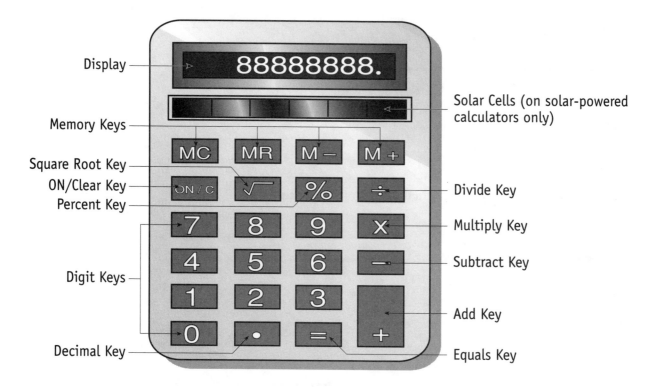

Locate the following keys (or similar keys) on your calculator.

- The **on/off key** [ON/OFF]. Press [ON/OFF] once to turn a calculator on and press it again to turn it off.

- The **digit keys** [0], [1], [2], [3], [4], [5], [6], [7], [8], and [9]. To enter a number, press one digit at a time.

- The **decimal point key** [·]. To enter dollars and cents or decimal numbers, press [·] to separate the whole number part from the decimal part.

- The **function keys** [+], [−], [×], and [÷]. Press a function key to add, subtract, multiply, or divide.

- The **clear key** [C]. Pressing [C] erases the display. Press [C] to begin a new problem or when you've made a keying error.

Other commonly used clear key symbols are

[ON/C] On/Clear   [CE/C] Clear Entry/Clear   [CE] Clear Entry   [AC] All Clear

Displayed Numbers

EXAMPLE 1   Enter 4,730 on your calculator.

Press Keys   **Display Reads**

[C] | 0. |

[4] [7] [3] [0] | 4730. |

Most calculators display zero and a decimal point when first turned on and after the clear key is pressed.

Most calculators display a decimal point to the right of a whole number.

You do not enter a comma to separate groups of digits.

EXAMPLE 2   Enter $5.94 on your calculator.

Press Keys   **Display Reads**

[C] | 0. |

[5] [.] [9] [4] | 5.94 |

You enter a decimal point to separate dollars from cents.

Calculator Keys
A calculator does not have a comma key (,) or a dollar sign key ($). On some calculators, a comma may be placed automatically by the calculator.

Use a calculator to answer the following questions.

1. With your calculator on, press the digit keys 1 through 9.

   [1] [2] [3] [4] [5] [6] [7] [8] [9]

   a. How many digits appear on your display? _____

   b. What is the largest number your calculator can display? _____

2. What keys do you press to clear the display? _____

3. What appears on the display after you push the clear key? _____

4. Enter the following on your calculator. Then show how the calculator displays each number or amount.

	Enter	Display Reads		Enter	Display Reads
a.	$0.58	_____	c.	239	_____
b.	$8.45	_____	d.	1,956	_____

Number Sense Review

Solve the problems below.

1. How is 2,043 written in words?

2. In 108, 619, what does 0 stand for?

3. How is three million, five hundred thousand, four hundred sixty written in digits?

4. Gina earns $6.45 per hour, and her friend Maria earns $7.12. Which expression comparing these two amounts is true?

    a. $6.45 > $7.12   **c.** $6.45 < $7.12   **e.** $7.12 = $6.45
    b. $6.45 = $7.12   **d.** $7.12 < $6.45

5. For which whole number is the following inequality true?

    5 × _____ > 34

    a. 3   **b.** 4   **c.** 5   **d.** 6   **e.** 7

6. The direct distance from New York to San Francisco is 3,267 miles. Round this distance to the nearest hundred miles.

7. Rounded to the nearest ten dollars, what is the price of the microwave oven at the right?

Better Homes
Microwave

$374.99

8. Estimate the total retail sales for the week shown at the right.

    a. $1,600   **d.** $2,200
    b. $1,800   **e.** $2,400
    c. $2,000

Day	Retail Sales
Monday	$379.85
Tuesday	$412.90
Wednesday	$406.25
Thursday	$387.40
Friday	$397.96
Total:	

9. Which of the following gives the best estimate of the purchase prices shown at the right?

$6.39 $2.07 $0.89

a.	$5.00	**b.**	$5.00	**c.**	$6.00	**d.**	$6.00	**e.**	$6.00
	3.00		2.00		2.00		3.00		3.00
	+ 1.00		+ 1.00		+ 1.00		+ 1.00		+ 2.00

For problems 10–13, use estimation to find each exact answer.

10. $3{,}087 + 2{,}943 =$

    a. 5,240   **b.** 5,660   **c.** 6,030   **d.** 6,420   **e.** 6,870

11. $823 - $297 =

    a. $526   **b.** $606   **c.** $716   **d.** $796   **e.** $846

12. $98 \times 21 =$

    a. 1,362   **b.** 1,572   **c.** 1,802   **d.** 2,058   **e.** 2,398

13. $7\overline{)}$ ^{6}

    a. 27   **b.** 82   **c.** 138   **d.** 175   **e.** 213

14. The lowest recorded temperature in the U.S. was –79°F. It occurred on January 23, 1971, in Prospect Creek, Alaska. About how many degrees below freezing (32°F) was that temperature?

15. Using the sales tax table, determine the amount of tax on a purchase of $19.16.

Amount of Sale	Tax
$18.75 to $18.91	$1.13
18.92 to 19.08	1.14
19.09 to 19.24	1.15
19.25 to 19.41	1.16
19.42 to 19.58	1.17

16. Which of the following lists the grains shown on the table in the order of calories—listing the grain with the most calories first, and so on?

    a. B, D, A, E, C   **d.** B, C, A, E, D
    b. D, B, C, E, A   **e.** D, B, E, A, C
    c. C, A, E, B, D

Grain (per cup)	Calories	Protein (g)
A. White Rice	186	3.7
B. Brown Rice	232	4.9
C. Oats	132	4.8
D. Rye	237	8.6
E. Wheat	221	4.8

WHOLE NUMBERS

Adding Whole Numbers

To add two or more numbers
- add from right to left, starting with the ones column
- regroup as needed

Use estimation or a calculator to check your answers.

> **Math Tip**
> When adding, **regrouping** is also called **carrying** or **renaming.**

EXAMPLE   Add.

STEP 1 Add the ones:
$8 + 5 = 13$
Write 3; regroup
10 ones as 1 ten.

STEP 2 Add the tens:
$1 + 7 + 9 = 17$
Write 7; regroup
10 tens as 1 hundred.

STEP 3 Add the hundreds:
$1 + 3 + 2 = 6$
Write 6. Check by
estimating.

```
  378        1            1 1          1 1
+ 295      3 7 8        3 7 8        3 7 8        400
           +2 9 5       +2 9 5       +2 9 5   is close to  + 300
             3            7 3        6 7 3  ————→    700
```

Calculator Check

Press Keys: [C] [3] [7] [8] [+] [2] [9] [5] [=]   (**Note:** A calculator displays an answer
Answer: [6 7 3 .   ]   only after you press [=].)

Add. Check by estimating.

1.
```
  ¹
  38      40     57        41        79        99
+ 19    + 20   + 38      + 26      + 74      + 35
  57      60
```

2.
```
  ¹
  272     300     329       490       626       533
+ 167   + 200   + 180     + 264     + 293     + 272
  439     500
```

3.
```
  ¹           ¹
  2,473     2,500    1,831      3,478     5,957     5,500
+   915   +   900   +   846    + 1,720   + 2,612   + 3,750
  3,388     3,400
```

4.
```
       ¹
  $6.75                                $4.36     $7.64     $8.19
+  3.19                              +  3.29    + 6.75    + 5.90
  $9.94
```
> Place a decimal point in the sum to separate dollars and cents.

Math Tip

Estimates can differ in value. When you are estimating by rounding, the value of an estimate depends on the place value to which the numbers are rounded.

Exact Answer	Estimate 1	Estimate 2	Estimate 3	
468	470	500	470	(nearest ten)
+ 209	+ 210	+ 200	+ 200	(nearest hundred)
677	680	700	670	
	Rounded to nearest ten	**Rounded to nearest hundred**	**Mixed rounding**	

As you can see, there are many ways to round to find an estimate.

Add. Check by estimating.

5.
```
    586        845        1,129      $1,853      2,493
  + 375      + 796      +   684     +    759   + 1,728
```

6.
```
    182       $347         808       3,500      $4,687
     75         89         437         753       2,158
  +  47      +   76      +   96      +   467    + 1,653
```

Add. As your first step, line up the digits: ones digit under ones digit, and so on.

7. 157 + 89 = $354 + $298 =   1,895 + 762 =

8. 1,308 + 875 = $3,375 + $2,836 =   12,654 + 8,682 =

Use addition to solve each word problem.

9. The distance from Seattle to Portland is 179 miles, and the distance from Portland to Eugene is 114 miles. Driving through Portland, how far is Eugene from Seattle?

10. Estimate the total cost of a $48 set of tools, a $33.88 wheelbarrow, and a $279.95 lawn mower. *Any reasonable estimate is acceptable.*

Subtracting Whole Numbers

To subtract two numbers
- subtract from right to left, starting with the ones column
- regroup as needed

Use estimation or a calculator to check your answers.

> **Math Tip**
> When subtracting, **regrouping** is also called **borrowing** or **renaming**.

EXAMPLE 1 Subtract.

$$\begin{array}{r} 74 \\ -39 \\ \hline \end{array}$$

STEP 1 Regroup 1 ten:

$$70 = 60 + 10$$

$$\begin{array}{r} 6 \\ \not7 4 \\ -3\ 9 \\ \hline \end{array}$$

STEP 2 Add 10 to the 4 ones:

$$10 + 4 = 14$$

$$\begin{array}{r} 6\ 14 \\ \not7\not4 \\ -3\ 9 \\ \hline \end{array}$$

STEP 3 Subtract:

$$14 - 9 = 5$$
$$6 - 3 = 3$$

$$\begin{array}{r} 6\ 14 \\ \not7\not4 \\ -3\ 9 \\ \hline 3\ 5 \end{array}$$

Check by estimating.

$$\begin{array}{r} 70 \\ \text{is close to} \quad -40 \\ \hline \longrightarrow \quad 30 \end{array}$$

Calculator Check

Press Keys: [C] [7] [4] [−] [3] [9] [=]
Answer: [35.]

(?) **Subtract. Check by estimating.**

1.
$$\begin{array}{r} 67 \\ -34 \\ \hline 33 \end{array} \qquad \begin{array}{r} 70 \\ -30 \\ \hline 40 \end{array} \qquad \begin{array}{r} 86 \\ -55 \\ \hline \end{array} \qquad \begin{array}{r} 67 \\ -57 \\ \hline \end{array} \qquad \begin{array}{r} 154 \\ -43 \\ \hline \end{array} \qquad \begin{array}{r} 286 \\ -75 \\ \hline \end{array}$$

2.
$$\begin{array}{r} {}^{2\ 14} \\ 3\not4 2 \\ -160 \\ \hline 182 \end{array} \qquad \begin{array}{r} 350 \\ -150 \\ \hline 200 \end{array} \qquad \begin{array}{r} 478 \\ -188 \\ \hline \end{array} \qquad \begin{array}{r} 549 \\ -264 \\ \hline \end{array} \qquad \begin{array}{r} 274 \\ -182 \\ \hline \end{array} \qquad \begin{array}{r} 856 \\ -376 \\ \hline \end{array}$$

3.
$$\begin{array}{r} {}^{4\ 18} \\ \$7.\not5\not8 \\ -3.49 \\ \hline \$4.09 \\ \uparrow \end{array}$$

Place a decimal point in the difference to separate dollars and cents.

$$\begin{array}{r} \$7.90 \\ -3.28 \\ \hline \end{array} \qquad \begin{array}{r} \$8.94 \\ -5.77 \\ \hline \end{array} \qquad \begin{array}{r} \$10.37 \\ -8.09 \\ \hline \end{array}$$

4.
$$\begin{array}{r} \$5.76 \\ -3.83 \\ \hline \end{array} \qquad \begin{array}{r} \$6.47 \\ -2.95 \\ \hline \end{array} \qquad \begin{array}{r} \$8.50 \\ -5.70 \\ \hline \end{array} \qquad \begin{array}{r} \$19.25 \\ -12.65 \\ \hline \end{array} \qquad \begin{array}{r} \$34.75 \\ -25.93 \\ \hline \end{array}$$

EXAMPLE 2 521 − 179

STEP 1
$$\begin{array}{r} 5\,\overset{1}{2}\,\overset{11}{\cancel{1}} \\ -1\,7\,9 \\ \hline \end{array}$$

STEP 2
$$\begin{array}{r} \overset{4}{\cancel{5}}\,\overset{11}{\cancel{2}}\,\overset{11}{\cancel{1}} \\ -1\,7\,9 \\ \hline 3\,4\,2 \end{array}$$

EXAMPLE 3 400 − 156

STEP 1
$$\begin{array}{r} \overset{3}{\cancel{4}}\,\overset{10}{\cancel{0}}\,0 \\ -1\,5\,6 \\ \hline \end{array}$$

STEP 2
$$\begin{array}{r} \overset{3}{\cancel{4}}\,\overset{9}{\overset{10}{\cancel{0}}}\,\overset{10}{\cancel{0}} \\ -1\,5\,6 \\ \hline 2\,4\,4 \end{array}$$

· ·

5.
$$\begin{array}{r} \overset{5}{\cancel{6}}\,\overset{16}{\cancel{7}}\,\overset{14}{\cancel{4}} \\ -2\,9\,8 \\ \hline 3\,7\,6 \end{array}$$
$$\begin{array}{r} 700 \\ -\,300 \\ \hline 400 \end{array}$$
$$\begin{array}{r} 375 \\ -176 \end{array}$$
$$\begin{array}{r} 203 \\ -145 \end{array}$$
$$\begin{array}{r} 634 \\ -276 \end{array}$$
$$\begin{array}{r} 748 \\ -389 \end{array}$$

6.
$$\begin{array}{r} 300 \\ -175 \end{array}$$
$$\begin{array}{r} 500 \\ -328 \end{array}$$
$$\begin{array}{r} 412 \\ -153 \end{array}$$
$$\begin{array}{r} 1{,}600 \\ -\;545 \end{array}$$
$$\begin{array}{r} 1{,}922 \\ -\;746 \end{array}$$

7.
$$\begin{array}{r} \$5.00 \\ -\;2.75 \end{array}$$
$$\begin{array}{r} \$6.00 \\ -\;3.85 \end{array}$$
$$\begin{array}{r} \$7.25 \\ -\;2.79 \end{array}$$
$$\begin{array}{r} \$14.00 \\ -\;6.39 \end{array}$$
$$\begin{array}{r} \$35.10 \\ -\;11.58 \end{array}$$

Subtract. As your first step, line up the digits.

8. 762 − 588 = $12.40 − $8.94 =   1,308 − 948 =

Use subtraction to solve each word problem.

9. The price of a $314 television is being reduced to $225 during the Memorial Day sale. How much can Gloria save by buying the set at this sale?

10. Find both an estimate and the exact change for Vernon's purchase. Vernon pays with a $50 bill.

$19.75

Estimate: _____

Exact Change: _____

11. Ervin weighs 79 pounds more than his wife, Rosey. What is Rosey's weight if Ervin tips the scale at 216 pounds?

Problem Solver: Understanding Set-Up Questions

Some test questions, called **set-up questions,** do not ask you to find the answer. Instead, a set-up question asks you to write or choose a numerical expression or an equation that shows how to compute the correct answer.

EXAMPLE 1 Shelley has driven 75 miles of the 340-mile distance between Springfield and Oakridge. Which expression shows how many miles she has left to drive?

a.	$340 + 75$	**d.**	$75 - 340$
b.	$75 + 340$	**e.**	$340 \div 75$
c.	$340 - 75$		

You think: How much is 340 take away 75?

You choose the expression that represents this subtraction.

The correct answer choice is **c. 340 – 75.**

EXAMPLE 2 For a 3-day weekend sale, the price of a toaster has been reduced by $5.89. The sale price is $17.60. Which equation tells how to calculate the original price p?

a.	$p = \$17.60 - \$5.89$	**d.**	$p = \$17.60 - \$11.71$
b.	$p = \$17.60 + \$5.89$	**e.**	$p = \$17.60 \times \$5.89$
c.	$p = \$17.60 + \$11.71$		

You think: The original price p is $17.60 plus $5.89.

You choose the equation that represents this addition.

The correct answer choice is **b. $p = \$17.60 + \$5.89.$**

As you see, solving a set-up question involves identifying a numerical expression that represents information given in a sentence or phrase. Here are some more examples.

Sentence or Phrase	Numerical Expression
What is $20.00 minus $14.95?	$\$20.00 - \$14.95$
How much smaller is 87 than 102?	$102 - 87$
What is the sum of 945 and 893?	$945 + 893$
By how much is $48 more than $31?	$\$48 - \$31$
124 feet decreased by 75 feet	$124 - 75$
the total of $8.46 and $5.76	$\$8.46 + \$5.76$
1,378 pounds increased by 479 pounds	$1,378 + 479$

Write a numerical expression for each sentence or phrase.

Sentence or Phrase	Numerical Expression

1. What is $25.00 minus $17.98? _____

2. Find the sum of 2,305 and 1,987. _____

3. How long is 62 inches added to 97 inches? _____

4. $1.98 taken away from $5.00 _____

5. $4.88 more than $2.19 _____

6. 1,345 pounds reduced by 277 pounds _____

Choose the correct expression or equation to solve each problem.

7. Jonas earned $75 on Monday, $86 on Tuesday, and $105 on Wednesday. Which expression shows the total amount Jonas earned for the three days?

   a. $105 − $86 − $75
   b. $105 + $86 − $75
   c. $105 + $86 + $75

8. Emma paid $1,850 less for her antique jewelry than she sold it for. If she sold it for $2,375, which equation represents the amount a that Emma paid for her antique jewelry?

   a. $a = \$2{,}375 + \$1{,}850$
   b. $a = \$2{,}375 - \$1{,}850$
   c. $a = \$1{,}850 - \$2{,}375$

9. In Hank's pocket are several coins: two quarters, three dimes, and four nickels. Which equation tells the total amount t of money, in cents, in Hank's pocket?

   a. $t = 25¢ + 30¢ + 20¢$
   b. $t = 50¢ + 40¢ + 15¢$
   c. $t = 50¢ + 30¢ + 20¢$

10. Carmen paid for a $58.88 wheelbarrow by cashing a paycheck for $203.64. Which expression gives the *best estimate* of the change Carmen should receive?

    a. $200 − $60
    b. $200 + $60
    c. $200 − $50

11. After driving 278 miles, Tuong sees a sign saying "Chicago: 412 miles." Which equation gives the *best estimate* of the total miles m Tuong will have driven by the time he reaches Chicago?

    a. $m = 400 - 300$
    b. $m = 400 + 200$
    c. $m = 400 + 300$

Problem Solver: Identifying Necessary Information

Many word problems (and problems involving tables, graphs, and maps) contain more information than you need to answer a specific question. For these problems, your first step is to decide which information is important.

Necessary information includes only those numbers needed to answer a specific question.

Extra information includes those numbers not needed to answer a specific question.

EXAMPLE 1   Henri hoped to spend no more than $35.00 for a new shirt and a new belt. If the shirt cost $19.95 and the belt cost $12.89, how much did Henri spend in all?

> **Necessary information:** $19.95, $12.89
>
> **Extra information:** $35.00

Item Sets

An **item set** is a group of two or more problems that are based on the same passage or graphic. For Examples 2 and 3, refer to the following passage.

> Woody is buying a used pickup that has a sticker price of $15,150. The dealer agreed to lower the price an extra $500 and to give Woody $6,500 for his old car as a trade-in. Woody will pay an additional down payment of $3,850. For the balance he still owes, Woody will take out a loan with the dealer. He will make 24 monthly payments of $195.50 each.

EXAMPLE 2   What price is Woody paying for the pickup before the amount of the trade-in is deducted?

> **Necessary information:** $15,150 and $500. The price is determined by subtracting $500 from $15,150.
>
> **Extra information:** $6,500, $3,850, 24, and $195.50

ANSWER: $14,650

EXAMPLE 3   How much of a loan will Woody take out with the dealer?

> **Necessary information:** $15,150, $500, $6,500, and $3,850. The loan amount is determined by subtracting each of the three amounts $500, $6,500, and $3,850 from the sticker price of $15,150.
>
> **Extra information:** 24, $195.50

ANSWER: $4,300

For problems 1–4, refer to the map at the right.

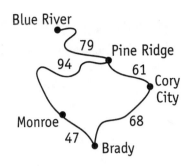

1. How many miles is Monroe from Cory City if you take the route through Brady?

2. How much farther is Pine Ridge from Monroe than Pine Ridge is from Cory City?

3. How far is Blue River from Brady if you take the route through Monroe?

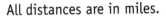

All distances are in miles.

4. To determine which of the two routes from Brady to Blue River is the shorter, which of the following distances is *not* needed?

   a. 47   b. 61   c. 68   d. 79   e. 94

For problems 5–8, refer to the following passage.

Leona is trying to decide where to take her daughter Amy for lunch. At Burger Village, children's burgers cost $1.19, french fries cost $0.89, and soft drinks cost $0.79. For herself, Leona would buy the Village Luncheon Special for $3.95. At The Sandwich Shop, a kid's meal costs $2.19 and includes a hamburger, french fries, and a soft drink. At The Sandwich Shop, Leona's meal costs $4.79.

5. If Amy has a hamburger, fries, and a soft drink, how much will Leona spend for Amy's meal at Burger Village?

6. If Amy has a kid's meal, which expression gives the *best estimate* of the total cost of lunch if Amy and Leona eat at The Sandwich Shop?

   a. $1.00 + $4.00   d. $3.00 + $5.00
   b. $2.00 + $4.00   e. $3.00 + $6.00
   c. $2.00 + $5.00

7. How much more would Leona spend on her own meal at The Sandwich Shop than at Burger Village?

8. Which amount is *not* needed to determine how much more Amy's meal costs at Burger Village than at The Sandwich Shop?

   a. $0.79   b. $0.89   c. $1.19   d. $2.19   e. $3.95

Multiplying Whole Numbers

To multiply two numbers
- multiply from right to left, starting with the ones
- regroup as needed
- write zeros as placeholders if you find it helpful when writing partial products

Use estimation or a calculator to check your answers.

> **Math Tip**
> **Remember:**
> - zero times a number is zero
> - one times a number is that number
> $7 \times 0 = 0$ $7 \times 1 = 7$

EXAMPLE Multiply.

STEP 1 Multiply 326 by 8.* Write 2608, writing 8 in the ones place.

STEP 2 Multiply 326 by 4.* Write 1304, writing 4 in the tens place.

STEP 3 Add the two partial products.

```
            2 4                        1 2
  3 2 6     3 2 6                      3 2 6              3 2 6
×   4 8   ×   4 8                   ×    4 8           ×    4 8
          first partial → 2 6 0 8       2 6 0 8           2 6 0 8
          product            second partial → 1 3 0 4 0 ←   1 3 0 4 0
                             product         placeholding   1 5,6 4 8
                                             zero
```

(***Note:** Each partial product has its own carried digits. It is a good idea to write carried digits lightly in pencil. Then you can erase them after you do each multiplication.)

Calculator Check

Press Keys: C 3 2 6 × 4 8 =
Answer: 15648.

....

Multiply. Check by estimating.

1.
23	56	38	51	74
× 3	× 5	× 7	× 7	× 9

2.
```
  4 3
$5.75          Place a decimal point in    $7.25      $12.50      $25.75
×   6          the product to separate    ×    8     ×    5      ×    9
$34.50         dollars and cents.
   ↑
```

3.
34	46	53	28	79
×25	×32	×47	×19	×65

For problems 4 and 5, a shortcut is shown that you can use when the bottom number contains a zero. Use a placeholding zero (as shown in the shortcut) instead of writing a partial product that contains only zeros (000).

Shortcut   Long Way

4.
$6.43
× 20
‾‾‾‾‾
$128.60

$6.43
× 20
‾‾‾‾‾
000
1286
‾‾‾‾‾
$128.60

$8.57
× 40
‾‾‾‾‾

739
× 60
‾‾‾‾‾

382
× 70
‾‾‾‾‾

Shortcut   Long Way

5.
864
× 605
‾‾‾‾‾
4320
518400
‾‾‾‾‾
522,720

864
× 605
‾‾‾‾‾
4320
0000
‾‾‾‾‾
518 400
‾‾‾‾‾
522,720

675
× 507
‾‾‾‾‾

396
× 306
‾‾‾‾‾

708
× 403
‾‾‾‾‾

Multiply. As your first step, line up the digits.

6. $79 \times 61 =$ $85 \times 43 =$ $\$5.67 \times 25 =$ $\$14.62 \times 50 =$

Use multiplication to solve each word problem.

7. What is the cost of 19 gallons of gasoline priced at $1.27 per gallon?

8. Find both an estimate and the exact price of the paint at the right.

   Estimate: _____

   Exact Price: _____

Wally buys 28 quarts at $10.39 per quart.

9. Kate's car gets 36 miles per gallon during highway driving. On a cross-country trip, how far can Kate's car go on a full tank of 18 gallons?

Dividing Whole Numbers

Dividing by a One-Digit Number

Division is a multistep process that combines many of the skills you have already studied. For longer problems, you may need to repeat these steps.

- Think of a long division problem as being made up of several short divisions.
- When division does not end in zero, write r and the amount left over to show the remainder.

> **Division Steps**
> - Divide; then write a digit in the quotient.
> - Multiply.
> - Subtract.
> - Compare.
> - Bring down the next digit from the dividend.

EXAMPLE 1 Divide. $4\overline{)95}$

First set of steps: Divide 4 into 9.

STEP 1 Divide: $9 \div 4 = 2$
Write 2 above the 9.

STEP 2 Multiply: $2 \times 4 = 8$

STEP 3 Subtract: $9 - 8 = 1$

STEP 4 Compare: $1 < 4$

STEP 5 Bring down the 5.

$$\begin{array}{r} 2 \\ 4\overline{)95} \\ -8\downarrow \\ \hline 15 \end{array}$$

Second set of steps: Divide 4 into 15.

STEP 1 Divide: $15 \div 4 = 3$
Write 3 above the 5.

STEP 2 Multiply: $3 \times 4 = 12$

STEP 3 Subtract: $15 - 12 = 3$

STEP 4 Compare: $3 < 4$

STEP 5 There is no other digit to bring down. Write 3 as the remainder.

$$\begin{array}{r} 23\,r\,3 \\ 4\overline{)95} \\ -8 \\ \hline 15 \\ -12 \\ \hline 3 \end{array}$$

Calculator Check

Press Keys: [C] [9] [5] [÷] [4] [=]
Answer: [23.75]

(**Note:** A calculator displays a remainder as a decimal fraction. You'll review decimals in a later chapter.)

Divide. Check by estimating.

1. $\begin{array}{r} 9\,r\,3 \\ 5\overline{)48} \\ -45 \\ \hline 3 \end{array}$ $\begin{array}{r} 10 \\ 5\overline{)50} \end{array}$ $6\overline{)25}$ $\qquad$ $7\overline{)54}$ $\qquad$ $3\overline{)19}$ $\qquad$ $8\overline{)60}$

2. $\begin{array}{r} 54 \\ 6\overline{)324} \\ -30 \\ \hline 24 \\ -24 \\ \hline 0 \end{array}$ $\begin{array}{r} 50 \\ 6\overline{)300} \end{array}$ $4\overline{)336}$ $\qquad$ $5\overline{)295}$ $\qquad$ $4\overline{)892}$ $\qquad$ $3\overline{)728}$

Using Zero as a Placeholder

Use a zero as a placeholder in the quotient each time you
- divide into a zero
- divide into a number that is smaller than the divisor

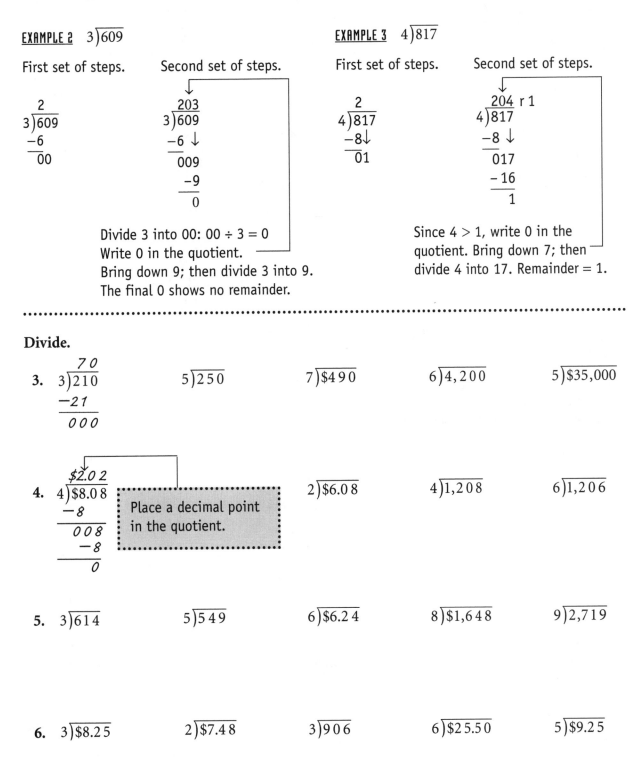

__EXAMPLE 2__ $3\overline{)609}$

First set of steps.

$$
\begin{array}{r}
2 \\
3\overline{)609} \\
-6 \\
\hline
00
\end{array}
$$

Second set of steps.

$$
\begin{array}{r}
203 \\
3\overline{)609} \\
-6\downarrow \\
\hline
009 \\
-9 \\
\hline
0
\end{array}
$$

Divide 3 into 00: 00 ÷ 3 = 0
Write 0 in the quotient.
Bring down 9; then divide 3 into 9.
The final 0 shows no remainder.

__EXAMPLE 3__ $4\overline{)817}$

First set of steps.

$$
\begin{array}{r}
2 \\
4\overline{)817} \\
-8\downarrow \\
\hline
01
\end{array}
$$

Second set of steps.

$$
\begin{array}{r}
204 \text{ r } 1 \\
4\overline{)817} \\
-8\downarrow \\
\hline
017 \\
-16 \\
\hline
1
\end{array}
$$

Since 4 > 1, write 0 in the
quotient. Bring down 7; then
divide 4 into 17. Remainder = 1.

....

Divide.

3.
$$
\begin{array}{r}
70 \\
3\overline{)210} \\
-21 \\
\hline
000
\end{array}
$$

$5\overline{)250}$ $7\overline{)\$490}$ $6\overline{)4,200}$ $5\overline{)\$35,000}$

4.
$$
\begin{array}{r}
\$2.02 \\
4\overline{)\$8.08} \\
-8 \\
\hline
008 \\
-8 \\
\hline
0
\end{array}
$$
Place a decimal point in the quotient.

$2\overline{)\$6.08}$ $4\overline{)1,208}$ $6\overline{)1,206}$

5. $3\overline{)614}$ $5\overline{)549}$ $6\overline{)\$6.24}$ $8\overline{)\$1,648}$ $9\overline{)2,719}$

6. $3\overline{)\$8.25}$ $2\overline{)\$7.48}$ $3\overline{)906}$ $6\overline{)\$25.50}$ $5\overline{)\$9.25}$

Dividing by a Two-Digit Number

You use the same multistep process, repeating the steps for each digit, to divide by a 2-digit number. Estimating can help with each division.

<u>EXAMPLE 4</u> Divide. $17\overline{)795}$

First set of steps: Divide 17 into 79.

STEP 1 Divide: $79 \div 17 = ?$
Estimate: $80 \div 20 = 4$
Try 4.

$$\begin{array}{r} 4 \\ 17\overline{)795} \\ -68 \\ \hline 115 \end{array}$$

STEP 2 Write 4 above 9.
Multiply: $17 \times 4 = 68$

STEP 3 Subtract: $79 - 68 = 11$

STEP 4 Compare: $11 < 17$
(4 is correct because $68 < 79$ and $11 < 17.^*$)

STEP 5 Bring down the 5.

Second set of steps: Divide 17 into 115.

STEP 1 Divide: $115 \div 17 = ?$
Estimate: $120 \div 20 = 6$
Try 6.

$$\begin{array}{r} 46 \text{ r } 13 \\ 17\overline{)795} \\ -68 \\ \hline 115 \\ -102 \\ \hline 13 \end{array}$$

STEP 2 Write 6 above 5.
Multiply: $17 \times 6 = 102$

STEP 3 Subtract: $115 - 102 = 13$

STEP 4 Compare: $13 < 17$
(6 is correct because $102 < 115$ and $13 < 17.^*$)

STEP 5 There is no other digit to bring down. Remainder = 13.

*If *both* comparison conditions are not met, choose a new quotient digit and try again.

....

**For problems 7 and 8, check to see if each quotient digit is correct.
Cross out each incorrect digit and write the correct digit above it.
Then complete the division problem.**

7. $18\overset{2}{\overline{)56}}$ $32\overset{3}{\overline{)97}}$ $24\overset{5}{\overline{)119}}$ $37\overset{5}{\overline{)219}}$ $28\overset{6}{\overline{)210}}$

8.
$$\begin{array}{r} \text{Check} \\ \downarrow \\ 18 \\ 28\overline{)549} \\ -28 \\ \hline 269 \end{array}$$
$$\begin{array}{r} \text{Check} \\ \downarrow \\ 23 \\ 19\overline{)422} \\ -38 \\ \hline 42 \end{array}$$
$$\begin{array}{r} \text{Check} \\ \downarrow \\ 24 \\ 33\overline{)823} \\ -66 \\ \hline 163 \end{array}$$
$$\begin{array}{r} \text{Check} \\ \downarrow \\ 45 \\ 16\overline{)750} \\ -64 \\ \hline 110 \end{array}$$
$$\begin{array}{r} \text{Check} \\ \downarrow \\ 32 \\ 22\overline{)718} \\ -66 \\ \hline 58 \end{array}$$

Divide. Check by estimating.

9. $27\overline{)243}$ $34\overline{)204}$ $18\overline{)144}$ $21\overline{)153}$ $38\overline{)347}$

10. $19\overline{)647}$ $31\overline{)589}$ $22\overline{)879}$ $43\overline{)907}$ $54\overline{)849}$

11. $15\overline{)1,650}$ $26\overline{)4,654}$ $35\overline{)8,295}$ $19\overline{)8,320}$ $46\overline{)12,675}$

Divide. As your first step, rewrite each problem with a division bracket.

12. $518 \div 7 =$ $605 \div 5 =$ $941 \div 8 =$ $1,654 \div 9 =$

13. $234 \div 18 =$ $452 \div 24 =$ $2,853 \div 13 =$ $6,050 \div 29 =$

Use division to solve each word problem.

14. If Yolanda can pack 36 calculators into each shipping box, how many boxes will she need to send 2,448 calculators?

15. Kirby Bus Company is busing 329 students to a music conference. If each bus holds a maximum of 38 students, how many buses will be needed?

Problem Solver: Solving Multistep Problems

In a multistep problem, you use two or more operations to compute an answer. The key to solving a multistep problem is to think of it as two or more one-step problems.

EXAMPLE 1   Maria and three friends agreed to share equally the cost of dinner. The bill included $21.75 for pizzas, $7.50 for salads, and $8.75 for soft drinks. How much was Maria's share of the bill?

You think: First, I need to know the total cost of the meal. Then, I divide this cost by 4 to find Maria's share.

STEP 1   Letting *c* stand for *cost of meal,* you can write
$c = \$21.75 + \$7.50 + \$8.75 = \$38.00$

STEP 2   Letting *M* stand for *Maria's share,* you write
$M = \$38.00 \div 4 = \mathbf{\$9.50}$

ANSWER: Maria's share is $9.50.

> **Math Tip**
> Although it's not necessary, writing an equation for each step is a good idea. Many people, though, prefer to write just the correct numerical expression.

In many multistep problems, there is more than one correct way to compute the answer.

EXAMPLE 2   Phil earns $7.50 for each hour he works on Thursday and Friday. How much will Phil earn on these 2 days when he works for 8 hours on Thursday and 6 hours on Friday?

You may think: First, I need to find how many total hours he'll work. Then, I multiply the total hours by $7.50.
If *h* stands for *total hours worked,* and *a* for *total amount earned,* you can write the two steps as

STEP 1 $h = 8 + 6 = 14$ hours

STEP 2 $a = \$7.50 \times 14 = \mathbf{\$105.00}$

Or you may think: First, I need to find out how much he earned each day. Then, I add the two amounts to find the total.
If *T* stands for *amount earned on Thursday,* *F* for *amount earned on Friday,* and *a* for *total amount earned,* you write

STEP 1 $T = \$7.50 \times 8$ and $F = \$7.50 \times 6$
          $= \$60.00$ $= \$45.00$

STEP 2 $a = \$60.00 + \$45.00 = \mathbf{\$105.00}$

ANSWER: Phil will earn $105.

Either solution gives the correct answer.

Write a numerical expression for each sentence or phrase.

Sentence or Phrase	Numerical Expression
1. What is the product of 123 times 17?	_____
2. How many times does 9 go into 236?	_____
3. What distance is twice 748 miles?	_____
4. $134.88 divided 6 ways	_____
5. the quotient 3,540 divided by 35	_____

Solve each word problem.

6. Herman baked cookies for his daughter's second-grade class. From the 93 cookies he baked, he kept 24 home for his family. He took the rest to school. If there are 23 children in the class, how many cookies will each child get?

7. Manny and three friends agreed to share the cost of dinner. They had a pizza for $14.70, soft drinks for $5.50, and salads for $8.40. What was Manny's share of the bill?

8. For her car tune-up business, Kira bought 12 cases of motor oil. Each case contains 24 quarts of oil. If each tune-up takes an average of 4 quarts, how many tune-ups can Kira do before she needs more oil?

9. As part of a promotion, Lita gave away 60 balloons every day during the 31 days in July. At the end of the month, she still had 132 balloons left over. Knowing this, figure out how many total balloons Lita started out with.

10. Roberto can carry 27 bales of hay in his pickup. He can carry an additional 15 bales in a small trailer attached to the pickup. How many bales can Roberto move in 7 trips if he uses both the pickup and the trailer?

11. Each Monday through Friday, Marla delivers a newspaper to each of her 74 customers. Last weekend she delivered a total of 114 additional papers. How many papers did Marla deliver last week?

12. Thomas planned to spend $35.00 at The Man's Store. After buying a shirt for $14.95 and a tie for $9.99, how much did Thomas have left to spend?

13. Look at the map below. How many more miles do you fly if you connect at Chicago on a trip from Denver to New York than if you fly direct?

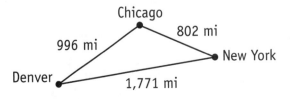

14. When Jerrie bought her Honda, the mileage indicator read 18,146 miles. After owning the car for 1 year, Jerrie notices that the indicator reads 33,062 miles. On the average, how many miles has Jerrie driven each month?

Problem Solver: Choosing the Correct Expression

In the previous problem solver, you saw how a multistep problem can be solved by breaking it down into simpler one-step problems. Each step has its own numerical expression.

On this page, you'll learn how to write a **single numerical expression** to represent all steps needed to solve a multistep problem. The key is to learn the correct order in which to solve these expressions.

Order of Operations

Parentheses ( ) are used to indicate a single quantity that is to be multiplied. When you use parentheses, you do not need to write a multiplication sign.

To find the value of a numerical expression, follow these steps:
- Perform any operation indicated within parentheses first.
- Next, perform any operations in the numerator or denominator of a fraction.
- Then, from left to right, perform all the multiplications and divisions.
- Last, from left to right, perform all the additions and subtractions.

Numerical Expression | | **Finding the Value**

$3 \times 4 + 9$ **STEP 1** Multiply: $3 \times 4 = 12$ **STEP 2** Add: $12 + 9 = \mathbf{21}$

$14 - 24 \div 8$ **STEP 1** Divide: $24 \div 8 = 3$ **STEP 2** Subtract: $14 - 3 = \mathbf{11}$

$\$20 - (\$5 + \$3)$ **STEP 1** Add: $\$5 + \$3 = \$8$ **STEP 2** Subtract: $\$20 - \$8 = \mathbf{\$12}$

$\dfrac{25 - 9}{8}$ **STEP 1** Subtract: $25 - 9 = 16$ **STEP 2** Divide: $16 \div 8 = \mathbf{2}$

$4(12 - 7)$ **STEP 1** Subtract: $12 - 7 = 5$ **STEP 2** Multiply: $4 \times 5 = \mathbf{20}$

$(\$19 + \$9) \div 7$ **STEP 1** Add: $\$19 + \$9 = \$28$ **STEP 2** Divide: $\$28 \div 7 = \mathbf{\$4}$

<u>EXAMPLE</u>   Jamie earns $9.75 for each hour she works on the weekend. If she worked for 8 hours on Saturday and 6 hours on Sunday, how much did she earn over the weekend? Write two numerical expressions that will solve the problem.

The two correct numerical expressions for the solution of this problem are
1. total earned = pay per hour times total hours worked
   = $\$9.75(8 + 6) = \mathbf{\$136.50}$
   or
2. total earned = Saturday's earnings plus Sunday's earnings
   = $(\$9.75 \times 8) + (\$9.75 \times 6) = \mathbf{\$136.50}$

Each numerical expression gives the same answer: **$136.50.**

Write a numerical expression for each sentence or phrase below.

Sentence or Phrase	**Numerical Expression**
1. How much is $18.99 added to the product of $5.50 times 7?	_____
2. Find the sum of 345 and 125; then divide the total by 3.	_____
3. What is $45 reduced by the quotient of $60 divided by 2?	_____
4. The sum of 37 plus 29 plus 14; the total multiplied by 4	_____
5. $100 divided by the sum of 2 + 4	_____

Find the value of each expression below.

6. $12 \times 4 - 16$ $9 \times 8 + 24$ $36 - 12 \div 2$

7. $\dfrac{35 - 19}{4}$ $5(15 - 9)$ $(\$24 + \$13) - (\$12 + \$9)$

For problems 8 and 9, circle the letter of the correct expression.

8. Amy gave a clerk $20.00 for a $10.98 gallon of paint and one paintbrush. The paint cost $10.98, the brush cost $4.89, and the sales tax was $0.88. Which expression shows how much change Amy should get?

   a. ($10.98 + $4.98 + $0.88) − $20.00   **d.** $20.00 − ($10.98 + $4.89 − $0.88)
   b. $20.00 − $10.98 − $4.89 + $0.88   **e.** $20.00 + $10.98 + $4.89 − $0.88
   c. $20.00 − ($10.98 + $4.89 + $0.88)

9. Sandi and two friends agreed to share the cost of lunch. Hamburgers cost $6.57, soft drinks $4.50, and salads $5.25. Which expression shows Sandi's share?

   a. ($6.57 + $4.50 + $5.25)2   **d.** $\dfrac{\$6.57 + \$4.50 + \$5.25}{3}$
   b. 3($6.57 + $4.50 + $5.25)   **e.** 2($6.57 + $4.50 + $5.25)
   c. $\dfrac{\$6.57}{3 + (\$4.50 + \$5.25)}$

For problem 10, circle two correct expressions for the answer.

10. Each day he works overtime, Daniel earns $18 more than his regular daily pay of $64. Last week, he worked overtime 2 out of 5 days. How much did Daniel earn last week?

   a. 3($64 × $18) + 2($64)   **d.** $64 × 3 + $18 × 2
   b. 5($64 + $18)   **e.** 2($64 + $18) + 3($64)
   c. $64 × 5 + $18 × 2

Data Highlight: Finding Typical Values

Look at the list of pizza prices shown at the right. Suppose that your friend Dana asks you, "What is the typical price of a small pizza in Albany?"

What would you tell Dana? $13, $14, $15, $16, or some other amount? Just what is *typical*?

In math, there are three main ways to give a **typical value:** as a **mean,** a **median,** or a **mode.**

Pizza Prices in Albany	
Joey's Small Pizza	$14
Mia's Small Pizza	$13
Tino's Small Pizza	$16
Little Italy's Small Pizza	$14
Franko's Small Pizza	$15

Mean

Mean is another word for **average.** To find the average of a set (group) of numbers

- add the numbers in the set
- divide this sum by the number of numbers in the set

Usually, when people talk about average, they're referring to mean.

(**Note:** In most cases, the mean is not equal to any number in a set. However, the mean is often close in value to the middle number of the set.)

EXAMPLE 1 To find the **mean** price of a small pizza in Albany, follow these steps.

STEP 1 Add.   **STEP 2** Divide.

$$
\begin{array}{r}
\$14 \\
13 \\
16 \\
14 \\
+\ 15 \\
\hline
\$72
\end{array}
$$

$14.40 ← mean
5)$72.00 ← sum
↑
number
in set

Mean = $14.40

Median

The **median** is the middle value of a set of numbers, arranged from least to greatest value.

- If a set contains an odd number of numbers, the median is the middle number.
- If a set contains an even number of numbers, the median is the average of the two middle numbers.

EXAMPLE 2 To find the **median** price of a small pizza in Albany, do the following.

STEP 1 Arrange the prices in order, from least to greatest.
$13, $14, $14, $15, $16
        └middle value

STEP 2 Since there is an odd number of prices, the median is the middle value.

Median = $14

Mode

The **mode** of a set of numbers is the number that appears the most times in the set. If no number appears more than once, a set has no mode.

EXAMPLE 3 The **mode** of the pizza prices is $14, since $14 appears more than any other price.

Mode = $14

For problems 1 and 2, find the mean, median, and mode.

1. Shoe prices

   $15 $18 $20 $26 $26

   Mean: _____

   Median: _____

   Mode: _____

2. Condominium sizes (square feet)

   1,400   1,400   1,550   1,850   2,000

   Mean: _____

   Median: _____

   Mode: _____

3. The four new models at Al's Car City have different mileage ratings.

   a. What is the average city mileage rating of the four models listed?

   b. What is the average highway mileage rating of the four models listed?

   c. What is the median city mileage rating of the four models listed? (**Remember:** For an even number, the median is the average of the two middle values.)

	City	Highway
Model DX	33	38
Model DXi	28	35
Model LX	26	31
Model SX	17	24

4. This chart shows prices of shoes being sold at Family Shoe Store.

   a. Compute the profit made on each model.

   b. Compute the mean profit of the four models listed.

   c. Determine the median profit of the four models listed.

Model	Selling Price (a)	Store's Cost (b)	Profit ($a-b$)
A	$60.00	$40.00	_____
B	$52.00	$36.00	_____
C	$48.00	$34.00	_____
D	$40.00	$25.00	_____

5. This chart lists Benita's overtime hours. Which equation shows how to compute the average weekly overtime hours (h) Benita worked during May?

   a. $h = (12 + 9 + 14) \div 3$
   b. $h = (9 + 8 + 12 + 10) \div 3$
   c. $h = (9 + 8 + 12 + 10) \div 4$
   d. $h = (12 + 9 + 14) \div 4$
   e. $h = 4(9 + 8 + 12 + 10)$

	April	May	June
Week 1	12	9	14
Week 2	10	8	12
Week 3	8	12	9
Week 4	10	10	13

Whole Numbers Review

Solve the problems below.

1. For July, Francine's employer withheld $278 for federal taxes, $59 for Social Security, and $37 for medical insurance. What is the total of these three deductions?

2. The first football game at Franklin High School had an attendance of 5,219. The second game drew only 4,847 fans. How much did the attendance decrease between these two games?

3. Erin's Restaurant sold 257 large colas over the weekend. If each large cup holds 16 ounces of soda, how many total ounces of cola did Erin's sell during this 2-day period?

4. For a loan of $5,000 from his cousin, Bert would have to pay back a total of $5,679.60 divided into 24 equal monthly payments. What would be Bert's monthly loan payment?

5. The sticker on Phan's new car rates its mileage at 36 miles per gallon for highway driving and 27 miles per gallon in the city. For city driving, how many miles can Phan drive if the tank holds 18 gallons?

6. Willem paid for a $28.99 shirt with a $50 bill. With the change, he's decided to buy as many $2 bottles of hair shampoo as he can. How many bottles is he able to buy?

7. Virginia wants to know how much more a set of six walnut chairs costs than a set of four pecan chairs and two maple chairs. Which amount is the *best estimate* of the difference?

Maple	Oak	Walnut	Pecan
$139.49	$131.29	$150.89	$121.79

  a. $60   **b.** $80   **c.** $100   **d.** $120   **e.** $140

8. A case (12 quarts) of motor oil normally sells for $11.99 but is on sale for $0.79 per quart. The oil company is offering a $1.50-per-case rebate. Suppose Alison buys a case at the sale price and sends for the rebate. Which expression shows how much the case will end up costing Alison?

   a. $11.99 − $1.50   **c.** $11.99 − $0.79 × 12   **e.** $0.79 × 12 − $1.50

   b. 12($1.50 − $0.79)   **d.** $11.99 − ($0.79 × 12 − $1.50)

9. Over the past 30 months, Bea's weight dropped from 230 to 148. Her goal was to lose 100 pounds and reach an ideal weight of 130. She figures this goal is still 6 months off. Up to now, though, which expression shows Bea's average monthly weight loss?

   a. $\dfrac{100}{(30 + 6)}$ **b.** $\dfrac{(230 - 148)}{30}$ **c.** $\dfrac{(230 - 130)}{30}$ **d.** $\dfrac{(230 - 130)}{(30 + 6)}$ **e.** $\dfrac{(230 - 148)}{(30 + 6)}$

For problems 10 and 11, refer to the following passage.

Monty applied for a job as a part-time administrative assistant with an accounting firm. As part of the application process, he had to take a 16-minute typing test. During the 16 minutes, he typed 1,072 words. Monty got the job at a starting salary of $1,388 per month. After 8 months, Monty got a raise to his present salary of $1,492 per month.

10. On the average, how many words per minute did Monty type during the typing test?

11. Which expression is the *best estimate* of Monty's earnings during his first 6 months on the job?

   a. $1,400 × 6   **c.** $1,400 × 8   **e.** $1,500 × 10

   b. $1,500 × 6   **d.** $1,500 × 8

12. What is the mean price of condominiums listed for the city of Lewisburg?

   a. $74,000   **d.** $86,540

   b. $80,900   **e.** $98,000

   c. $83,080

13. Which equation shows how to compute the median price of Marysville condominiums?

   a. $p = (\$82{,}500 + \$87{,}000) \div 6$

   b. $p = (\$72{,}000 + \$97{,}650) \div 6$

   c. $p = (\$82{,}500 + \$87{,}000) \div 6$

   d. $p = (\$72{,}000 + \$97{,}650) \div 2$

   e. $p = (\$82{,}500 + \$87{,}000) \div 2$

Condominiums For Sale

Lewisburg	Garrett	Marysville
$68,500	$74,800	$72,000
$74,000	$78,000	$76,500
$80,900	$86,900	$82,500
$94,000	$92,000	$87,000
$98,000	$99,900	$91,800
	$104,000	$97,650

NUMBERS LESS THAN 1

Parts of a Whole

You can think of a number less than 1 as part of a whole. People often refer to parts of a whole:

"Drive three tenths of a mile and then turn right."

"You'll need a quarter pound of butter for the cake."

"Forty percent of the class was absent on Tuesday."

You can divide any whole object, such as a pound or a mile, or any group, such as a class, into smaller parts.

If you represent the whole object with the number 1, you can represent any part of that object with a number less than 1.

In this chapter, you'll learn the meanings and uses of three ways to write numbers less than 1.

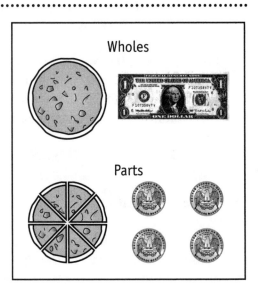

- decimal fractions: 0.25 *or* twenty-five hundredths

- proper fractions: $\frac{1}{4}$ *or* one fourth *or* one quarter

- percents: 25% *or* 25 percent *or* twenty-five percent

Listed below are commonly used words that stand for parts of a whole. Be sure to know the meaning and spelling of each of these words.

1. Write the correct words in the Skill Check column to name the part shown.

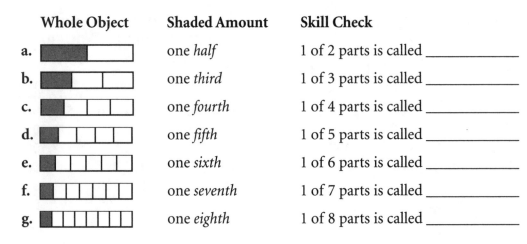

Whole Object	Shaded Amount	Skill Check
a.	one *half*	1 of 2 parts is called _____
b.	one *third*	1 of 3 parts is called _____
c.	one *fourth*	1 of 4 parts is called _____
d.	one *fifth*	1 of 5 parts is called _____
e.	one *sixth*	1 of 6 parts is called _____
f.	one *seventh*	1 of 7 parts is called _____
g.	one *eighth*	1 of 8 parts is called _____

Whole Object	**Shaded Amount**	**Skill Check**
h.	one *ninth*	1 of 9 parts is called _____
i.	one *tenth*	1 of 10 parts is called _____
j.	one *hundredth*	1 of 100 parts is called _____
100		
k.	one *thousandth*	1 of 1,000 parts is called _____
1,000		

2. For each statement below, circle the whole that is being discussed.
   Then underline the part.

   a. "Kent spent almost half an hour on the telephone!"

   b. "The bank will loan us 90 percent of the $14,500 we need for the car."

   c. "The recipe calls for three fourths of a pound of beef."

   d. "Our train station is only four tenths of a mile away."

   e. "The blouse is on sale for 35 percent off the regular price."

   f. "The machinist shaved eight hundredths of an inch off the fitting."

3. Write *d, f,* or *p* to indicate whether each of the following is usually
   expressed as a decimal (*d*), fraction (*f*), or percent (*p*).

   _____ **a.** a number of cents less than one dollar

   _____ **b.** the length of a bolt that's shorter than 1 inch

   _____ **c.** the amount of discount to be given at a store sale

   _____ **d.** the rate of interest that a bank offers on a savings account

   _____ **e.** human body temperature as it's read on a thermometer

   _____ **f.** a small amount of flour that is measured in a measuring cup

Reading and Writing Decimals

The First Three Decimal Places

The first three decimal places are **tenths, hundredths,** and **thousandths.** Most decimals that you use in daily life have three or fewer digits.

tenths ⌐ hundredths ⌐ thousandths

Math Tip	
Decimal	**Meaning**
0.1	1 of 10 equal parts
0.01	1 of 100 equal parts
0.001	1 of 1,000 equal parts
↑	

It is common to write a **leading zero** if the whole number is 0.

Reading Decimals

- To read a decimal, first read the number. Then say the place value of the right-most digit in the decimal number.

- Ignore the leading 0 when reading a decimal.

Example	Number + Place Value
0.1	**one tenth**
0.5	five tenths
0.01	**one hundredth**
0.05	five hundredths
0.23	twenty-three hundredths
0.001	**one thousandth**
0.005	five thousandths
0.023	twenty-three thousandths
0.150	one hundred fifty thousandths

Math Tip
Read a decimal point as the word *and*. Do not use the word *and* anywhere else when you say a number.
$7.29 is "seven dollars *and* twenty-nine cents"
127.52 is "one hundred twenty-seven *and* fifty-two hundredths"

Write each decimal in words.

1. 0.3

2. 0.5

3. 0.09

4. 0.20

5. 0.007

6. 0.080

7. 0.075

8. 0.425

Write each amount in words. Write the decimal point as *and*.

9. $2.05

10. $12.07

11. $3.88

12. $40.50

Writing Decimals

To write a decimal
- identify the place value of the right-hand digit
- write the number so that the right-hand digit is in its proper place. Then write zeros as placeholders if necessary.

EXAMPLE 1 Write 27 hundredths as a decimal.

Write 27 so that the 7 ends up in the hundredths place. (The hundredths place is the second place to the right of the decimal point.)

Write a leading zero.
↓
0.27
↑
Write 7 in the hundredths place.

EXAMPLE 2 Write 48 thousandths as a decimal.

Write 48 so that the 8 ends up in the thousandths place. (The thousandths place is the third place to the right of the decimal point.)

Write 0 as a placeholder.
↓
0.048
↑
Write 8 in the thousandths place.

Write each number as a decimal on the form at the right.

13. Fill in each amount on the purchase order.

Type A, fifty-six thousandths of an inch.
Type B, three tenths of an inch.
Type C, twenty-four thousandths of an inch.
Type D, two hundred nine thousandths of an inch.
Type E, one hundred twenty thousandths of an inch.
Type F, seven hundredths of an inch.
Type G, eight tenths of an inch.

Purchase Order

Item Type	Wall Thickness
Type A	_____ inch
Type B	_____ inch
Type C	_____ inch
Type D	_____ inch
Type E	_____ inch
Type F	_____ inch
Type G	_____ inch

14. Reporting on a 100-meter race, the announcer told the crowd by how much each runner "just missed the track record."

Lopez, seventy-eight thousandths of a second.
Jenkins, two tenths of a second.
Morris, fifty-five hundredths of a second.
Myers, one hundred twelve thousandths of a second.
Calder, six tenths of a second.
Nomura, eighty-six hundredths of a second.

Race Results

Runner	Time Off Record
Lopez	_____ sec
Jenkins	_____ sec
Morris	_____ sec
Myers	_____ sec
Calder	_____ sec
Nomura	_____ sec

Comparing and Ordering Decimals

A **mixed decimal** is a whole number plus a decimal. When you are comparing decimals and mixed decimals, use the familiar comparison symbols at the right (first discussed on page 24).

To make comparison easier, add zeros to give decimals the same number of decimal places.

Writing a zero at the right-hand end of a decimal does not change its value.

Math Tip
> means *is greater than*
< means *is less than*
= means *is equal to*
≠ means *is not equal to*

0.5 = 0.50

5 tenths = 50 hundredths

Place a zero at the end
0.5 = 0.50
↑
└── like this

not next to the decimal point.
0.5 ≠ 0.05
↑
└── not like this

EXAMPLE 1 Compare 0.35 and 0.42

0.$\boxed{3}$2   4 > 3, so
0.$\boxed{4}$2   0.42 > 0.32
↑
└── different first digits

EXAMPLE 2 Compare $2.65 and $2.70

$2.$\boxed{6}$5   6 < 7, so
$2.$\boxed{7}$0 $2.65 < $2.70
↑
└── same first digit

EXAMPLE 3 Compare 0.18 and 0.187

Add a 0. ┐
0.18$\boxed{0}$ ← 0 < 7, so
0.18$\boxed{7}$ 0.18 < 0.187
↑↑↑
└── same first two digits

Write >, <, or = to compare each pair of decimals.

1. 0.6 _____ 0.8   0.93 _____ 0.9   0.13 _____ 0.24

2. $0.09 _____ $0.15 $0.18 _____ $0.70 $0.40 _____ $.40

3. 3.56 _____ 3.359   1.50 _____ 1.5   12.456 _____ 12.63

4. $2.45 _____ $1.98 $0.89 _____ $1.09 $12.45 _____ $9.99

List the swimmers in order of finish. (Remember: The shortest time wins in a race.)

Race #1 Times	Race #1 Results		Race #2 Times	Race #2 Results
5. Bill: 48.8 sec	1st: _____	**6.**	Jackie: 24.75 sec	1st: _____
James: 49.08 sec	2nd: _____		Marina: 23.8 sec	2nd: _____
Pablo: 48.79 sec	3rd: _____		Fran: 23.785 sec	3rd: _____
Jamal: 49.125 sec	4th: _____		Anna: 24.29 sec	4th: _____

Life Skill: Reading a Metric Ruler

One of the most commonly used measuring tools is the **ruler.**
Pictured below is a **15-centimeter ruler.** Each centimeter (cm) is divided
into 10 **millimeters** (mm): 1 cm = 10 mm.

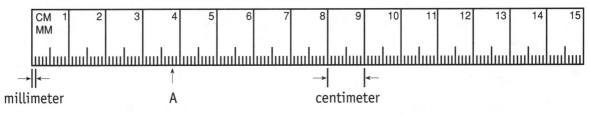

millimeter   A   centimeter

You read a distance on a centimeter ruler in several ways. Two of the
ways are

- as centimeters *and* millimeters: Point A is read as **3 cm 8 mm.**

- as centimeters *only:* Point A is read as **3.8 cm.**

EXAMPLE 1 Write 5 cm 7 mm as centimeters only.

EXAMPLE 2 Write 8.2 cm as centimeters and millimeters.

EXAMPLE 3 Write 49 mm in two other ways.

ANSWER: 5.7 cm

ANSWER: 8 cm 2 mm

ANSWER: 4 cm 9 mm or 4.9 cm

Write each length as indicated.

1. Write 9 cm 6 mm as centimeters only.

2. Write 3.4 cm as centimeters and millimeters.

3. Write 138 mm in two other ways.

    cm and mm: _____
    cm only: _____

4. What is the length of each object pictured below? Write your answers on the lines provided.

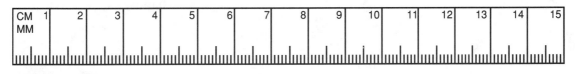

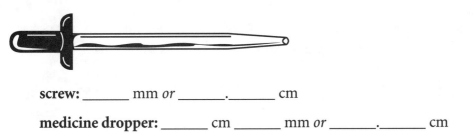

screw: _____ mm *or* _____ . _____ cm

medicine dropper: _____ cm _____ mm *or* _____ . _____ cm

Writing Proper Fractions

A **proper fraction,** such as $\frac{2}{3}$, $\frac{1}{8}$, or $\frac{15}{18}$, represents a number less than 1. (An **improper fraction** represents a number greater than or equal to 1, such as $\frac{6}{6}$, $\frac{7}{3}$, or $\frac{10}{1}$.) In any fraction

- the top number (the **numerator**) tells the number of parts you're describing
- the bottom number (the **denominator**) tells the number of equal parts the whole is divided into

The whole may be a single object   *or*   **the whole may be a group of objects.**

EXAMPLE 1 How much of the circle is shaded?

$\frac{3}{4}$ ← numerator
    ← denominator

3 parts are shaded.
The circle is divided
into 4 equal parts.

Read $\frac{3}{4}$ as three fourths.

EXAMPLE 2 What fraction of the group of circles is shaded?

Four out of five circles are shaded.
$\frac{4}{5}$ ← circles shaded
    ← circles in all
Read $\frac{4}{5}$ as four fifths.

For problems 1–5, write the fraction *and* the word name to show how much of each figure or group is shaded.

1.

_____ or _____

2.
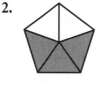

_____ or _____

3.

_____ or _____

4.
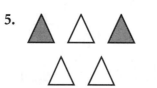

_____ or _____

5.

_____ or _____

6. Shade $\frac{2}{3}$ or this circle.

7. Shade $\frac{3}{5}$ of the group of squares below.

8. Shade $\frac{7}{8}$ of the distance between 0 and 1.

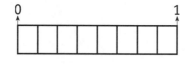

9. What fraction of a dollar is shown below?

Picturing Equal Fractions

There is more than one way to write a proper fraction to represent a given amount.

For example, the two cups contain equal amounts of syrup.

Equal Amounts

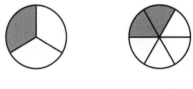

$\frac{6}{8}$ full = $\frac{3}{4}$ full

Equal Fractions

- The cup at the left is divided into 8 equal measuring units.

- The cup at the right is divided into 4 equal measuring units.

The fractions $\frac{3}{4}$ and $\frac{6}{8}$ represent the same amount. The two fractions are called **equal fractions** (or **equivalent fractions**).

Write equal fractions for each pair of figures as indicated.

1. the shaded fraction of each rectangle

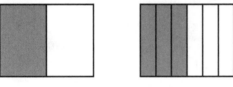

$$\frac{1}{2} = \frac{}{6}$$

2. the shaded fraction of each circle

$$\frac{1}{3} = \frac{}{}$$

3. the eaten fraction of each pizza

$$\frac{}{} = \frac{}{}$$

4. the shaded fraction of each group

$$\frac{}{} = \frac{}{}$$

5. the fraction of water in each cup

$$\frac{}{} = \frac{}{}$$

6. the length of the bar as a fraction of an inch

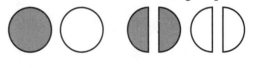

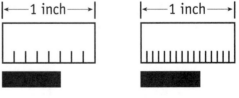

$$\frac{}{} = \frac{}{}$$

Reducing Fractions

To **reduce** or **simplify** a fraction is to rewrite it as an equal fraction with smaller numbers. You can reduce a fraction by dividing both the numerator and the denominator by the same number. When a fraction is in its simplest form—smallest numbers possible—it is said to be **reduced to lowest terms.**

<u>EXAMPLE</u>

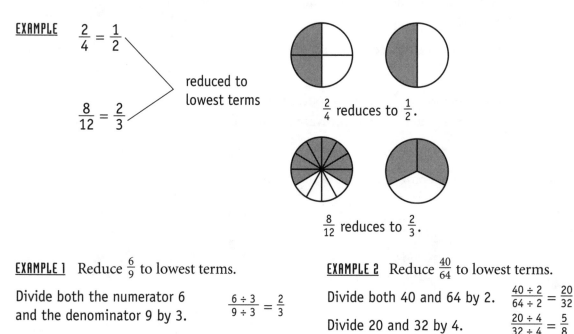

$$\frac{2}{4} = \frac{1}{2}$$

$$\frac{8}{12} = \frac{2}{3}$$

reduced to lowest terms

$\frac{2}{4}$ reduces to $\frac{1}{2}$.

$\frac{8}{12}$ reduces to $\frac{2}{3}$.

<u>EXAMPLE 1</u> Reduce $\frac{6}{9}$ to lowest terms.

Divide both the numerator 6 and the denominator 9 by 3.

$$\frac{6 \div 3}{9 \div 3} = \frac{2}{3}$$

<u>EXAMPLE 2</u> Reduce $\frac{40}{64}$ to lowest terms.

Divide both 40 and 64 by 2.

$$\frac{40 \div 2}{64 \div 2} = \frac{20}{32}$$

Divide 20 and 32 by 4.

$$\frac{20 \div 4}{32 \div 4} = \frac{5}{8}$$

The fraction $\frac{40}{64}$ could have been reduced to lowest terms in a single step by dividing by 8. $\frac{40}{64} = \frac{40 \div 8}{64 \div 8} = \frac{5}{8}$

Notice that 8 is the product of the two numbers that were used.

Reduce each fraction to lowest terms.

1. $\frac{3}{9} =$ $\qquad$ $\frac{2}{8} =$ $\qquad$ $\frac{4}{6} =$ $\qquad$ $\frac{6}{9} =$ $\qquad$ $\frac{6}{8} =$

2. $\frac{4}{10} =$ $\qquad$ $\frac{8}{12} =$ $\qquad$ $\frac{9}{15} =$ $\qquad$ $\frac{10}{14} =$ $\qquad$ $\frac{12}{16} =$

3. $\frac{15}{45} =$ $\qquad$ $\frac{14}{16} =$ $\qquad$ $\frac{8}{32} =$ $\qquad$ $\frac{10}{24} =$ $\qquad$ $\frac{28}{64} =$

Writing an Amount as a Fraction of a Larger Unit

Sometimes in a measurement problem you have to write one amount as a fraction of a larger unit.

EXAMPLE 1 What fraction of a yard is 32 inches?
(1 yard = 36 inches)

Write: $\dfrac{32}{36}$ $\leftarrow$ $\dfrac{\text{part}}{\text{whole}}$

Reduce: $\dfrac{32}{36} = \dfrac{32 \div 4}{36 \div 4} = \dfrac{8}{9}$

ANSWER: 32 inches is $\dfrac{8}{9}$ of a yard.

EXAMPLE 2 What fraction of an hour is
40 minutes? (1 hour = 60 minutes)

Write: $\dfrac{40}{60}$ $\leftarrow$ $\dfrac{\text{part}}{\text{whole}}$

Reduce: $\dfrac{40}{60} = \dfrac{40 \div 20}{60 \div 20} = \dfrac{2}{3}$

ANSWER: 40 minutes is $\dfrac{2}{3}$ of an hour.

Write each amount as a fraction. Reduce to lowest terms.

1. 1 yard = 36 inches

   a. 12 inches = _____ yard

   b. 24 inches = _____ yard

   c. 27 inches = _____ yard

2. 1 meter = 100 centimeters

   a. 20 centimeters = _____ meter

   b. 50 centimeters = _____ meter

   c. 60 centimeters = _____ meter

3. 1 pound = 16 ounces

   a. 6 ounces = _____ pound

   b. 10 ounces = _____ pound

   c. 12 ounces = _____ pound

4. 1 hour = 60 minutes

   a. 15 minutes = _____ hour

   b. 20 minutes = _____ hour

   c. 45 minutes = _____ hour

5. 1 year = 52 weeks

   a. 4 weeks = _____ year

   b. 20 weeks = _____ year

   c. 32 weeks = _____ year

6. 1 cup = 8 fluid ounces

   a. 4 fluid ounces = _____ cup

   b. 6 fluid ounces = _____ cup

   c. 8 fluid ounces = _____ cup

Solve each problem. Reduce fractions to lowest terms.

7. Ellen pays $900 each month for rent. If Ellen's monthly salary is $3,150, what fraction of her salary is needed to pay rent?

8. Miguel has delivered 12,000 pounds of an ordered 64,000 pounds of rock to the construction site. What fraction of the total needed has he delivered?

Raising Fractions to Higher Terms

You **raise a fraction to higher terms** by rewriting it as an equal fraction with larger numbers.

You often raise fractions to higher terms when you compare fractions and when you add or subtract fractions.

To raise a fraction to higher terms, *multiply* both the numerator and the denominator by the same number.

Equal Fractions

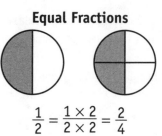

$$\frac{1}{2} = \frac{1 \times 2}{2 \times 2} = \frac{2}{4}$$

EXAMPLE   Write $\frac{3}{4}$ as a fraction that has 20 as a denominator.

STEP 1   Look at the denominators 4 and 20.
Ask, "What do I multiply 4 by to get 20?
$4 \times 5 = 20$

To Solve
$$\frac{3}{4} = \frac{?}{20}$$

STEP 2   Multiply the numerator 3 by 5: $3 \times 5 = 15$
The new numerator is 15.

Think
$$\frac{3 \times 5}{4 \times 5} = \frac{15}{20}$$

ANSWER: $\frac{3}{4} = \frac{15}{20}$

Raise each fraction to higher terms.

1. $\overset{\times 4}{\frac{1}{2}} = \frac{}{8}$ $\;\times 4$
    $\qquad \frac{1}{3} = \frac{}{6}$
    $\qquad \frac{1}{4} = \frac{}{8}$
    $\qquad \frac{2}{3} = \frac{}{9}$
    $\qquad \frac{3}{4} = \frac{}{12}$

2. $\frac{2}{5} = \frac{}{15}$
    $\qquad \frac{3}{4} = \frac{}{8}$
    $\qquad \frac{2}{3} = \frac{}{12}$
    $\qquad \frac{5}{7} = \frac{}{14}$
    $\qquad \frac{2}{6} = \frac{}{18}$

3. $\overset{\times 5}{\frac{2}{3}} = \frac{}{15}$ $\;\times 5$
    $\qquad \frac{2}{5} = \frac{}{30}$
    $\qquad \frac{5}{14} = \frac{}{28}$
    $\qquad \frac{11}{12} = \frac{}{36}$
    $\qquad \frac{4}{7} = \frac{}{28}$

4. $\frac{5}{7} = \frac{}{42}$
    $\qquad \frac{5}{6} = \frac{}{24}$
    $\qquad \frac{7}{8} = \frac{}{24}$
    $\qquad \frac{3}{4} = \frac{}{36}$
    $\qquad \frac{5}{8} = \frac{}{24}$

Comparing Proper Fractions

Like fractions are fractions that have the same denominator—called a **common denominator.** To compare like fractions, compare the numerators. The fraction with the larger numerator is the larger fraction.

Unlike fractions have different denominators. To compare unlike fractions, rewrite them as like fractions.

- First, choose a common denominator. Often, the larger denominator of the compared fractions can be chosen.
- Next, rewrite each fraction as an equal fraction that has this common denominator.
- Finally, compare the numerators of the like fractions.

Comparing **Like** Fractions

$\frac{3}{4}$ is greater than $\frac{2}{4}$

Comparing **Unlike** Fractions

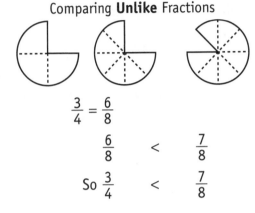

$$\frac{3}{4} = \frac{6}{8}$$

$$\frac{6}{8} \quad < \quad \frac{7}{8}$$

So $\frac{3}{4} \quad < \quad \frac{7}{8}$

EXAMPLE 1

Like Fractions

Compare $\frac{3}{4}$ and $\frac{2}{4}$.

In the numerators, 3 > 2.

So $\frac{3}{4} > \frac{2}{4}$.

EXAMPLE 2

Unlike Fractions

Compare $\frac{3}{4}$ and $\frac{7}{8}$.

STEP 1 Choose 8 as the common denominator.

STEP 2 Write $\frac{3}{4}$ as an equal fraction having a denominator of 8.

$$\frac{3}{4} = \frac{3 \times 2}{4 \times 2} = \frac{6}{8}$$

STEP 3 Compare $\frac{6}{8}$ and $\frac{7}{8}$.

In the numerators, 6 < 7. So $\frac{6}{8} < \frac{7}{8}$ or $\frac{3}{4} < \frac{7}{8}$.

Write >, <, or = to compare each pair of fractions.

1. $\frac{3}{6}$ ——— $\frac{2}{6}$ $\frac{5}{8}$ ——— $\frac{7}{8}$ $\frac{15}{16}$ ——— $\frac{13}{16}$ $\frac{19}{32}$ ——— $\frac{23}{32}$

Rewrite unlike fractions as like fractions before comparing. Use the larger denominator in each pair as the common denominator.

2. $\frac{2}{3}$ ——— $\frac{7}{12}$ $\frac{1}{4}$ ——— $\frac{3}{12}$ $\frac{2}{5}$ ——— $\frac{3}{10}$ $\frac{2}{3}$ ——— $\frac{5}{6}$

3. $\frac{3}{4}$ ——— $\frac{7}{8}$ $\frac{2}{3}$ ——— $\frac{8}{9}$ $\frac{4}{7}$ ——— $\frac{9}{14}$ $\frac{2}{5}$ ——— $\frac{6}{15}$

Finding the Lowest Common Denominator (LCD)

To compare unlike fractions, you rewrite them as like fractions. The key step is choosing a **common denominator.**

What is a common denominator for $\frac{2}{3}$ and $\frac{3}{4}$?

A common denominator is a number that each denominator divides into evenly. The smallest possible common denominator is called the **lowest common denominator** (LCD).

To find the lowest common denominator for $\frac{2}{3}$ and $\frac{3}{4}$ (and for other problems involving unlike fractions), use the **method of comparing multiples.**

EXAMPLE   Compare the fractions $\frac{2}{3}$ and $\frac{3}{4}$.

STEP 1   Write several multiples of each denominator.

Multiples of 3:   3   6   9   12   15
Multiples of 4:   4   8   12   16   20

STEP 2   Choose the smallest number that is on both lists: 12
12 is the LCD.

STEP 3   Rewrite each fraction as an equal fraction with a denominator of 12.

$$\frac{2}{3} = \frac{2 \times 4}{3 \times 4} = \frac{8}{12} \qquad \frac{3}{4} = \frac{3 \times 3}{4 \times 3} = \frac{9}{12}$$

STEP 4   Compare the like fractions and compare the original fractions.

$$\frac{8}{12} < \frac{9}{12}$$

So $\frac{2}{3} < \frac{3}{4}$.

Write the first five multiples of each of the following numbers.

1. 2: *2, 4, 6, 8, 10*   3:   4:   5:

2. 6:   8:   10:   12:

Use the method of comparing multiples to find the LCD for each pair of fractions.

3. $\frac{1}{4}$ and $\frac{1}{3}$ $\frac{2}{3}$ and $\frac{1}{5}$ $\frac{1}{2}$ and $\frac{1}{3}$ $\frac{3}{8}$ and $\frac{1}{6}$

   4: *4, 8, 12, 16*   3:   2:   8:

   3: *3, 6, 9, 12*   5:   3:   6:

   LCD: LCD: LCD: LCD:

Math Tip

Another way to find a common denominator is to multiply denominators by one another.

Although this method always gives you a common denominator you can use, it does *not* always give you the lowest common denominator.

At the right, both 48 and 24 can be used as common denominators, but 24 is smaller. It is the lowest common denominator, and it is easier to work with.

Find a common denominator for $\frac{1}{6}$ and $\frac{1}{8}$.

By Multiplying Denominators	By Comparing Multiples
$\frac{1}{6} = \frac{8}{48}$	$\frac{1}{6} = \frac{4}{24}$
$\frac{1}{8} = \frac{6}{48}$	$\frac{1}{8} = \frac{3}{24}$

Write >, <, or = to compare each pair of fractions below.

STEP 1　Determine the LCD.

STEP 2　Use the LCD to rewrite each fraction.

STEP 3　Compare the like fractions.

4.　$\frac{2}{3}$ _____ $\frac{4}{6}$　　　$\frac{1}{4}$ _____ $\frac{1}{3}$　　　$\frac{2}{3}$ _____ $\frac{2}{5}$　　　$\frac{2}{3}$ _____ $\frac{3}{4}$

5.　$\frac{1}{2}$ _____ $\frac{3}{5}$　　　$\frac{7}{8}$ _____ $\frac{3}{4}$　　　$\frac{2}{3}$ _____ $\frac{1}{2}$　　　$\frac{2}{3}$ _____ $\frac{5}{7}$

Arrange each group of three fractions in order, from the least to the greatest.

STEP 1　Use the largest denominator in each group as the LCD.

STEP 2　Use the LCD to rewrite each fraction.

STEP 3　Compare the like fractions.

6.　$\frac{2}{3}$　$\frac{7}{12}$　$\frac{3}{4}$　　　　$\frac{2}{3}$　$\frac{7}{9}$　$\frac{13}{18}$　　　　$\frac{3}{8}$　$\frac{1}{3}$　$\frac{7}{24}$　　　　$\frac{6}{14}$　$\frac{4}{7}$　$\frac{1}{2}$

Writing Improper Fractions and Mixed Numbers

In an **improper fraction,** the numerator is either greater than or equal to the denominator.

$\frac{4}{4}$ The top 4 stands for the number of pieces you're referring to. The bottom 4 stands for the number of pieces in 1 whole.

Thus, the value of $\frac{4}{4}$ is 1.

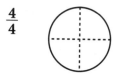

$\frac{7}{4}$ One whole is divided into 4 pieces. The numerator refers to 7 pieces. This is 3 pieces more than 1 whole.

Thus, the value of $\frac{7}{4}$ is greater than 1.

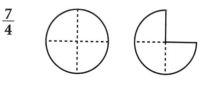

A **mixed number** is a whole number together with a fraction.

The whole number 1 stands for one whole object.

Along with the 1 object, there is $\frac{3}{4}$ of an object more.

Notice that $\frac{7}{4}$ and $1\frac{3}{4}$ represent the same amount. An amount greater than 1 can be written as either an improper fraction or a mixed number.

Write either an improper fraction or a mixed number to represent each amount below.

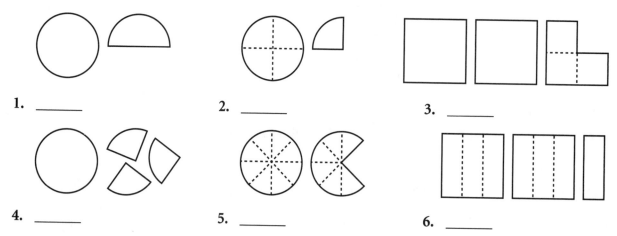

1. _____ 2. _____ 3. _____

4. _____ 5. _____ 6. _____

For problems 7 and 8, write your answer as an improper fraction and as a mixed number.

7. While doing inventory, Martin counted 9 quarter-pound cubes of butter. How much butter is this?

8. Glenna cuts pies into 8 equal pieces and then sells them by the slice. How many pies did Glenna sell if she sold 19 slices?

Life Skill: Reading an Inch Ruler

Pictured below is a 6-inch ruler.

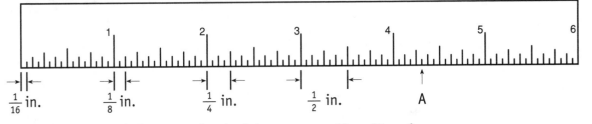

On the ruler, each fraction of an inch is represented by a line of different height. For example, the smallest markings represent $\frac{1}{16}$ of an inch.

EXAMPLE How far is point A from the left end of the ruler?

First notice that point A is between 4 and 5 inches from the left end.

Because point A is at a line that represent sixteenths of an inch, count how many sixteenths from 4 inches to point A.

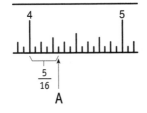

ANSWER: $4\frac{5}{16}$ **inches or** $4\frac{5}{16}''$ **(The symbol " stands for inches.)**

Write each measurement as an equivalent measurement using sixteenths of an inch.

1. **a.** $\frac{1}{8}$ in. = $\frac{}{16}$ in.   **b.** $\frac{1}{4}$ in. = $\frac{}{16}$ in.   **c.** $\frac{1}{2}$ in. = $\frac{}{16}$ in.   **d.** 1 in. = $\frac{}{16}$ in.

What distance is represented in each ruler pictured below?

2. **a.**   **b.**   **c.**   **d.**

_____ inch   _____ inch   _____ inch   _____ inch

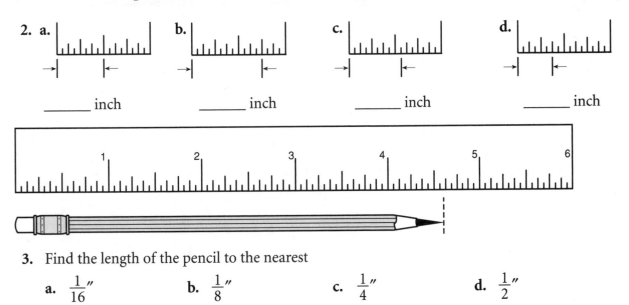

3. Find the length of the pencil to the nearest

   a. $\frac{1}{16}''$ **b.** $\frac{1}{8}''$ **c.** $\frac{1}{4}''$ **d.** $\frac{1}{2}''$

Understanding Percent

You can write part of a whole as a percent.

Percent refers to *the number of parts out of 100.*

is 1% of

You write percent as the *number of hundredths* followed by the percent sign %.

- A 5% sales tax means that you must pay $0.05 tax for every dollar of your purchase price.
- A 4% savings rate means that $100 in savings will earn $4 in interest each year.

Dividing a whole into 100 equal parts is one way to visualize percent.

Each large square is divided into 100 smaller squares.

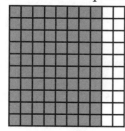

25% of the large square is shaded.

80% of the large square is shaded.

A dollar is equal to 100 cents.

75% of a dollar is shown.

What is 100%?

100% stands for a whole object. $100\% = 1$

At the right, 35% of the large square is shaded; 65% is unshaded.

$35\% + 65\% = 100\%$

shaded + unshaded = whole square

What Does a Percent Greater than 100% Mean?

A percent greater than 100% represents a number greater than 1.

200% stands for 2 whole objects, or twice an amount.
300% stands for 3 whole objects, or 3 times an amount, and so on.

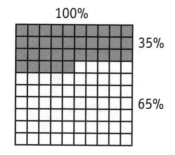

Math Tip
A sales increase from $100 to $300 is a **200% increase.**

Amount of Increase
$300 − $100 = **$200**

Percent of Increase
$\frac{\text{Increase}}{\text{Original amount}} = \frac{\$200}{\$100} = 2$ or 200%

For problems 1–4, write the percent of each square that is shaded, the percent that is unshaded, and the total of the two percents.

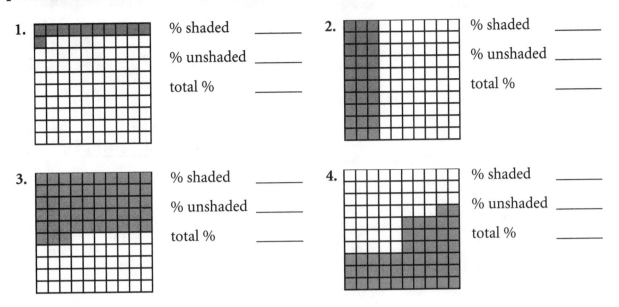

1. % shaded _____
    % unshaded _____
    total % _____

2. % shaded _____
    % unshaded _____
    total % _____

3. % shaded _____
    % unshaded _____
    total % _____

4. % shaded _____
    % unshaded _____
    total % _____

Solve each problem.

5. Chase County is trying to raise $100,000 in this year's United Way Fund Drive.
   a. If the fundraisers hope to raise 40% of the total during the first 2 weeks, how much will they need to collect during these 14 days?
   b. After collecting $85,000, what percent of the goal will they have reached?

6. Some of the country's best drivers race in the Kingston Classic car race. The race is 100 laps around a 2-mile-long track.
   a. What percent of the race does Mario have remaining after he's completed 42 laps?
   b. After driving 150 miles, what percent of the race has Mario finished?
   (**Hint:** $\frac{150}{200} = \frac{?}{100}$)

7. About 100,000 people attended last year's Rose Bowl game. The announcer said, "Had seats been available, we could have sold 400,000 tickets!"
   a. If the announcer was right, how many additional tickets could have been sold?
   b. What percent increase in sales do these additional tickets represent?

8. Last year, Brett Furniture had sales of $500,000. This year it had sales of $2,500,000.
   a. By how much money did Brett's sales increase from last year to this year? Express your answer in dollars.
   b. By what percent did Brett's sales increase from last year to this year?

Relating Decimals, Fractions, and Percents

A percent can easily be written as an equivalent decimal or fraction.

- Percent has the same value as a two-place decimal.
  37% is equal to **0.37**.

- Percent has the same value as a fraction that has a denominator of 100.
  37% is equal to $\frac{37}{100}$.

100 Equal Parts

37% is shaded.

0.37 is shaded.

$\frac{37}{100}$ is shaded.

> **Math Tip**
> Here is one way to remember the meaning of percent.
> - To write a percent as a decimal, let the two zeros in % remind you of two decimal places: 37% = 0.37.
> - To write the percent as a fraction, let the two zeros in % remind you of two zeros in the denominator: $\frac{37}{100}$.

Write each percent as an equivalent decimal.

1. 25%   37%   50%   78%   98%

Write each percent as an equivalent whole number.

2. 300%   600%   200%   400%   500%

Write each percent as an equivalent fraction. Reduce each fraction when possible.

3. 20%   25%   50%   60%   75%

Write each percent shown below as a decimal (*d*) and as a fraction (*f*).

4. Save 20%

   *d:*_____ *f:*_____

5. Savings Rate
   Now 5%

   | 5% Savings |

   *d:*_____ *f:*_____

6. New Home Loans
   30-Year Fixed Rate 8%

   *d:*_____ *f:*_____

Determine whether each number sentence is true or false. Circle T if the statement is true or F if it is false.

7. $25\% < \frac{1}{2}$ T or F $50\% > \frac{3}{4}$ T or F $20\% = \frac{2}{5}$ T or F

8. $40\% < \frac{1}{3}$ T or F $70\% \neq 0.07$ T or F $200\% > 0.250$ T or F

Answer each question. Reduce fractions to lowest terms.

9. Brent rode in a 100-mile bike race.

    a. What percent of the race had he finished after riding 60 miles?
    b. At this point, what *fraction* of the race did he still have left?

10. A ton is equal to 2,000 pounds.

    a. What fraction of a ton is 1,500 pounds?
    b. What percent of a ton is 1,500 pounds?
        (**Hint:** $\frac{1,500}{2,000} = \frac{?}{100}$)

11. By 9:30, Aaron had completed $\frac{3}{5}$ of his homework.

    a. By 9:30, what percent of his homework had Aaron completed?
        (**Hint:** $\frac{3}{5} = \frac{?}{100}$)
    b. How do you write $\frac{3}{5}$ as a decimal?

12. Jimmy went to a sale where "all trousers are marked down 25%."

    a. Express this price reduction as a decimal.
    b. Express this price reduction as a fraction.

13. A **meter** (a little longer than a yard) is a metric length unit that is equal to 100 centimeters.

    a. Expressed as a fraction, what part of a meter is 90 centimeters?
    b. Expressed as a percent, what part of a meter is 30 centimeters?
    c. Expressed as a decimal, what part of a meter is a length of 63 centimeters?

14. A **kilometer** (about 0.6 mile) is a metric length unit that is equal to 1,000 meters. What part of a kilometer is 450 meters?

    a. as a decimal:_____
        (**Hint:** 1 meter = 0.001 kilometer)
    b. as a fraction reduced to hundredths:_____
        (**Hint:** $\frac{450}{1,000} = \frac{?}{100}$)
    c. as a percent:_____

Data Highlight: Reading a Circle Graph

A common way to graphically display parts of a whole is by using a **circle graph.**

> **Math Tip**
> Because it resembles a cut pie, a circle graph is often called a **pie graph** or **pie chart.**

- An entire circle represents a whole amount.

- Individual sections represent parts of the whole, the largest section representing the largest portion, and so on.

On a circle graph, the sum of all sections adds up to 1 or 100%.

EXAMPLE   The Monroe family budget of $45,000 is divided into six expense categories. Using a circle graph, you can display the Monroe budget in any of three ways.

Graph A. Decimals*
(Each expense is displayed as *cents per budget dollar.*)

Graph B. Fractions
(Each expense is displayed as a *fraction of the total.*)

Graph C. Percents
(Each expense is displayed as a *percent of the total.*)

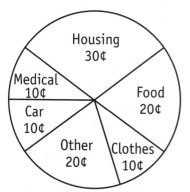 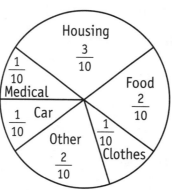 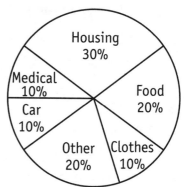

*In a circle graph, an amount such as thirty cents can be written as 30¢ or as $0.30.

Graph A: How many more cents per dollar do the Monroes spend on food than on clothes?

ANSWER: 20¢ – 10¢ = **10¢**

Graph B: Which is the largest expense category of the Monroe family?

ANSWER: Housing ($\frac{3}{10}$ is the largest fraction shown.)

Graph C: What total percent of the Monroe family budget is spent on medical, food, and housing?

ANSWER: 10% + 20% + 30% = **60%**

For problems 1–3, refer to Graph D.

1. How many more cents per dollar of sales does the QRT Boat Company make from North American sales than from European sales?

2. What number of cents should be written in the section labeled Other?

3. Are Pacific Rim sales double the amount of European sales?

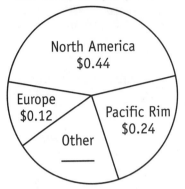

**Graph D
QRT Boat Company Sales**

North America
$0.44

Europe
$0.12

Pacific Rim
$0.24

Other
____

Cents per dollar of sales

For problems 4–6, refer to Graph E.

4. Write the fractions indicated on the preference poll as fractions all having a common denominator.

   Flats: $\frac{1}{6} =$ Pumps: $\frac{1}{4} =$ Sandals: $\frac{1}{6} =$

   Thongs: $\frac{1}{12} =$ Tennis Shoes: $\frac{1}{3} =$

5. According to the results of the preference poll, which type of shoe is most preferred by the women who were surveyed?

6. Which fraction is the *mode* of the results found in the preference poll?

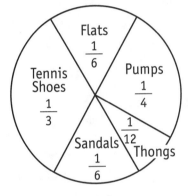

**Graph E
Women's Shoes
Preference Poll**

Flats
$\frac{1}{6}$

Pumps
$\frac{1}{4}$

Tennis
Shoes
$\frac{1}{3}$

Sandals
$\frac{1}{6}$

$\frac{1}{12}$
Thongs

For problems 7–9, refer to Graph F.

7. What total percent of their monthly take-home income do the Smiths spend on rent, food, and car payment?

8. Of the specific expense categories listed (not counting Other), which of the five percents is the *median?*

9. What percent of the Smith's family take-home income should be written in the section labeled Other?

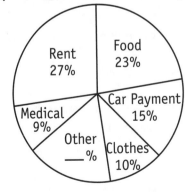

**Graph F
Smith Family Monthly Budget
(monthly take-home income)**

Rent
27%

Food
23%

Car Payment
15%

Medical
9%

Other
___%

Clothes
10%

Numbers Less than 1 Review

Solve the problems below.

1. Write the calculator display in words.

   | 0.002 |

2. Kim was told that the hole she was to drill in a piece of sheet metal should be about thirty-two hundredths of an inch. Write this number in digits.

3. Which of the following inequalities is *not* true?

   a. $0.8 > 0.79$ **b.** $0.14 < 0.23$ **c.** $1.3 > 1.25$ **d.** $2.5 < 1.99$ **e.** $1.4 > 0.75$

4. In a 100-meter dash, the following times were recorded:

   Al: 11.34 sec   Jess: 10.82 sec
   Bill: 10.7 sec   Bobby: 11.09 sec

   Listing the winner first, what is their correct finishing order?

   a. Bill, Jess, Bobby, Al   **c.** Al, Bobby, Jess, Bill   **e.** Jess, Bill, Bobby, Al
   b. Jess, Bobby, Bill, Al   **d.** Bobby, Jess, Al, Bill

5. Students in the Mr. Alvarez's class measured this key and gave four measurements. Which two measurements are correct?

   a. 68 mm   **c.** 6.08 cm
   b. 608 mm   **d.** 6.8 cm

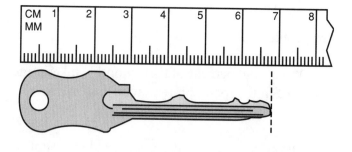

6. Forty minutes can be written as a fraction of an hour. How is this fraction written in its most reduced form?

7. What is the lowest common denominator for the fractions $\frac{3}{4}$ and $\frac{5}{6}$?

8. Judy has been asked to list the following three board thicknesses in order of size:

   $\frac{7}{12}$ inch

   Board A

   $\frac{2}{3}$ inch

   Board B

   $\frac{3}{4}$ inch

   Board C

   Listing the *narrowest board first,* what is the correct order of sizes?

   a. A, C, B   **b.** A, B, C   **c.** B, A, C   **d.** B, C, A   **e.** C, A, B

9. A number greater than 1 can be written as a mixed number *or* as an improper fraction. Which of the following does *not* correctly represent the amount of pizza shown?

   a. $\frac{12}{8}$ **c.** $\frac{8}{12}$ **e.** $\frac{6}{4}$

   b. $1\frac{4}{8}$ **d.** $1\frac{1}{2}$

10. To the nearest sixteenth of an inch, what is the length of the nail shown?

11. Barney baked 100 pastries for a school picnic. Of these 100 pastries, 65 were doughnuts and the rest were croissants. What percent of Barney's pastries were croissants?

12. A kilogram (about 2.2 pounds) is a metric unit of weight that's equal to 1,000 grams. Expressed as a percent, what part of a kilogram is a weight of 250 grams?

13. What percent of sales of the Thompson Shoe Company is made in the United States?

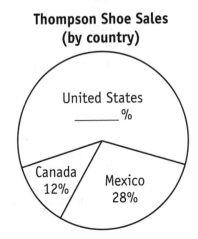

Thompson Shoe Sales (by country)

United States ____ %

Canada 12%

Mexico 28%

FRACTIONS

Building Confidence by Estimating

Some people feel unsure of their answers when working with fractions. Why is this? Here's a typical response:

"I just don't have a feeling about how answers should turn out."

If you've had a similar feeling, estimation can help. You can use estimation to find answers, check answers, and help choose among answer choices on a test.

As a first step, learn to recognize when a fraction is close to 0, $\frac{1}{2}$, or 1.

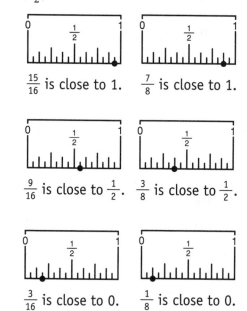

- A fraction is close to 1 when the denominator and numerator are about the same size.

$\frac{15}{16}$ is close to 1.　　$\frac{7}{8}$ is close to 1.

- A fraction is close to $\frac{1}{2}$ when the denominator is about twice as large as the numerator.

$\frac{9}{16}$ is close to $\frac{1}{2}$.　　$\frac{3}{8}$ is close to $\frac{1}{2}$.

- A fraction is close to 0 when the numerator is very small compared to the denominator.

$\frac{3}{16}$ is close to 0.　　$\frac{1}{8}$ is close to 0.

Estimating to Add Proper Fractions

Estimating is especially useful when you are adding proper fractions. The first step is to round each fraction.

- Round a fraction to 1 if the denominator and numerator are about equal.
  Fractions that round to 1: $\frac{3}{4}$, $\frac{5}{6}$, $\frac{7}{8}$, $\frac{9}{10}$, $\frac{11}{12}$, $\frac{15}{16}$

- Round a fraction to $\frac{1}{2}$ if the denominator is about twice the numerator.
  Fractions that round to $\frac{1}{2}$: $\frac{1}{3}$, $\frac{2}{3}$, $\frac{3}{5}$, $\frac{3}{8}$, $\frac{5}{9}$, $\frac{7}{12}$

- Round a fraction to 0 if the numerator is much smaller than the denominator.
  Fractions that round to 0: $\frac{1}{5}$, $\frac{1}{6}$, $\frac{2}{7}$, $\frac{1}{8}$, $\frac{1}{10}$, $\frac{1}{12}$, $\frac{3}{16}$

> **Math Tip**
> A fraction that's close to two different values can be rounded to either one! Either estimate will work fine. Remember to allow yourself flexibility when estimating.

When you are estimating a sum of proper fractions, group two halves $(\frac{1}{2})$ to make a whole (1). Then add the ones, the halves, and the zeros.

Estimate the following sums.

EXAMPLE 1

$$\frac{3}{8} \to \frac{1}{2}$$
$$\frac{4}{9} \to \frac{1}{2} \Big\rangle 1$$
$$+\frac{6}{7} \to 1$$

Estimate: 2

EXAMPLE 2

$$\frac{4}{7} \to \frac{1}{2}$$
$$\frac{3}{6} \to \frac{1}{2} \Big\rangle 1$$
$$+\frac{1}{8} \to 0$$

Estimate: 1

EXAMPLE 3

$$\frac{11}{12} \to 1$$
$$\frac{7}{8} \to 1$$
$$\frac{9}{10} \to 1$$
$$+\frac{1}{2} \to \frac{1}{2}$$

Estimate: $3\frac{1}{2}$

EXAMPLE 4

$$\frac{15}{16} \to 1$$
$$\frac{1}{16} \to 0$$
$$\frac{3}{8} \to \frac{1}{2}$$
$$+\frac{5}{6} \to 1$$

Estimate: $2\frac{1}{2}$

Estimate a sum for each problem by rounding the fractions to 1, $\frac{1}{2}$, or 0.

1.
$$\frac{2}{4}$$
$$\frac{7}{8}$$
$$+\frac{1}{6}$$

$$\frac{7}{8}$$
$$\frac{1}{5}$$
$$+\frac{2}{3}$$

$$\frac{8}{9}$$
$$\frac{3}{4}$$
$$+\frac{2}{5}$$

$$\frac{5}{6}$$
$$\frac{1}{2}$$
$$+\frac{9}{10}$$

$$\frac{15}{16}$$
$$\frac{3}{7}$$
$$+\frac{1}{3}$$

Estimate a solution to each problem below.

2. The three spacers are to be placed end to end. What will be their approximate combined length?

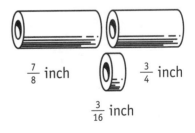

$\frac{7}{8}$ inch $\frac{3}{4}$ inch

$\frac{3}{16}$ inch

3. Every Monday, Wednesday, and Friday, Lydia swims $\frac{1}{2}$ mile. Tuesdays and Thursdays she swims $\frac{7}{8}$ mile. About how many miles does Lydia swim each week?

Estimating with Mixed Numbers

To estimate with mixed numbers, round each mixed number to the nearest whole number.

- If the value of the fraction is greater than or equal to $\frac{1}{2}$, round to the next higher whole number.
- If the value of the fraction is less than $\frac{1}{2}$, drop the fraction.

<u>**EXAMPLE 1**</u> Estimate: $14\frac{11}{12} - 7\frac{1}{3}$

STEP 1 Round each mixed number to the nearest whole number.

$14\frac{11}{12} \rightarrow 15$ because $\frac{11}{12} > \frac{1}{2}$

$7\frac{1}{3} \rightarrow 7$ because $\frac{1}{3} < \frac{1}{2}$

STEP 2 Subtract: $15 - 7 = 8$

ANSWER: $14\frac{11}{12} - 7\frac{1}{3} \approx 8$

<u>**EXAMPLE 2**</u> Estimate: $9\frac{1}{4} \times 7\frac{2}{3}$

STEP 1 Round each mixed number to the nearest whole number.

$9\frac{1}{4} \rightarrow 9$ because $\frac{1}{4} < \frac{1}{2}$

$7\frac{2}{3} \rightarrow 8$ because $\frac{2}{3} > \frac{1}{2}$

STEP 2 Multiply: $9 \times 8 = 72$

ANSWER: $9\frac{1}{4} \times 7\frac{2}{3} \approx 72$

(**Note:** The $\approx$ symbol means "is approximately equal to.")

Estimate an answer for each problem below by rounding mixed numbers to whole numbers

4.
$$9\frac{2}{3}$$
$$+\,7\frac{1}{4}$$

$$7\frac{1}{8}$$
$$+\,3\frac{2}{3}$$

$$16\frac{7}{10}$$
$$+\quad 9$$

$$24\frac{1}{2}$$
$$+\,17\frac{8}{9}$$

$$32\frac{11}{16}$$
$$+\,21\frac{1}{3}$$

5.
$$12\frac{1}{2}$$
$$-\,\,9\frac{7}{8}$$

$$9\frac{7}{8}$$
$$-\,3\frac{1}{4}$$

$$23\frac{15}{16}$$
$$-\,\,8\frac{1}{16}$$

$$18\frac{1}{10}$$
$$-\,12$$

$$11\frac{5}{12}$$
$$-\,\,6\frac{7}{8}$$

6. $8\frac{3}{4} \times 7\frac{1}{2} \approx$ $\qquad$ $5\frac{1}{3} \times 4\frac{4}{5} \approx$ $\qquad$ $14 \times 10\frac{1}{4} \approx$ $\qquad$ $20\frac{5}{8} \times 8\frac{1}{5} \approx$

To estimate when dividing, round to compatible whole numbers. Then divide.

7. $14\frac{1}{2} \div 3\frac{7}{8} \approx$ $\qquad$ $23\frac{2}{3} \div 8\frac{1}{4} \approx$ $\qquad$ $73\frac{7}{8} \div 9\frac{1}{4} \approx$ $\qquad$ $65\frac{1}{3} \div 9\frac{1}{2} \approx$

(?) Find a reasonable estimate for each problem below.

8. While at the deli, Linda bought $1\frac{3}{4}$ pounds of sliced turkey, $2\frac{7}{8}$ pounds of sliced beef, and $4\frac{3}{16}$ pounds of sliced ham. About how much meat did Linda buy in all?

9. Linda's husband, Mario, bought $1\frac{7}{8}$ pounds of cat food for $3.89 per pound and $2\frac{1}{8}$ pounds of dog food for $2.19 per pound. Approximately how much did Mario's purchase cost?

10. During the weekend storm, $5\frac{1}{4}$ inches of snow fell on Saturday and $4\frac{7}{8}$ inches on Sunday. An additional $3\frac{1}{2}$ inches fell on Monday. About how much snow fell on these three days?

11. Losing weight at the rate of $3\frac{1}{4}$ pounds per month, approximately how many months will it take Yoshi to lose a total of $28\frac{3}{4}$ pounds?

12. A one-cubic-foot container can hold $7\frac{1}{2}$ gallons of liquid. About how many gallons of gasoline does it take to fill a gas tank that has a volume of $3\frac{1}{8}$ cubic feet?

13. The value of TAQ stock slid last week from $56\frac{1}{4}$ to $38\frac{7}{8}$ following news of a major lawsuit against the company. By about how much did the value of TAQ stock drop last week?

14. From a piece of oak molding that is $32\frac{1}{2}$ inches long, Sue plans to cut off two pieces: one measuring $14\frac{1}{4}$ inches and one $9\frac{7}{8}$ inches. Which expression below gives the best estimate of how many inches of molding Sue will have left after cutting off these two pieces?

   a. $33 - 14 + 10$ **c.** $32 - 14 - 9$ **e.** $(14 - 10) + 33$
   b. $33 - (14 + 10)$ **d.** $(14 + 10) - 33$

15. For Halloween, Timothy bought a $15\frac{1}{4}$-pound bag of candy chews at bulk rate. He took out $2\frac{1}{8}$ pounds for his family and then divided the rest into 50 treat bags. Which expression gives the best estimate of how much candy will be in each treat bag?

   a. $\frac{50}{15-2}$ lb   **b.** $\frac{50}{15+2}$ lb   **c.** $\frac{15-2}{50+17}$ lb   **d.** $\frac{15+2}{50}$ lb   **e.** $\frac{15-2}{50}$ lb

Changing an Improper Fraction to a Mixed Number

The result of a calculation often is an improper fraction. Here are the steps to rewrite an improper fraction as a mixed number.

- Divide the denominator into the numerator.
- Write the remainder as a fraction.

EXAMPLE 1 Rewrite $\frac{14}{3}$ as a mixed number.

STEP 1 Divide 3 into 14.

$$\begin{array}{r} 4\ r\ 2 \\ 3\overline{)14} \\ -12 \\ \hline 2 \end{array}$$

STEP 2 The divisor is 3, so write the quotient 4 r 2 as the mixed number $4\frac{2}{3}$.

ANSWER: $4\frac{2}{3}$

EXAMPLE 2 Rewrite $\frac{12}{8}$ as a mixed number.

STEP 1 Divide 8 into 12.

$$\begin{array}{r} 1\ r\ 4 \\ 8\overline{)12} \\ -\ 8 \\ \hline 4 \end{array}$$

STEP 2 The divisor is 8, so write the quotient 1 r 4 as $1\frac{4}{8}$. Reduce the fractional part of the mixed number.

$$\frac{4 \div 4}{8 \div 4} = \frac{1}{2}$$

ANSWER: $1\frac{1}{2}$

Rewrite each improper fraction as a mixed number.

1. $\frac{7}{3} =$ $\frac{13}{10} =$ $\frac{9}{2} =$ $\frac{11}{3} =$ $\frac{14}{5} =$

2. $\frac{9}{6} =$ $\frac{18}{4} =$ $\frac{14}{4} =$ $\frac{23}{5} =$ $\frac{26}{6} =$

3. At his deli, Sal gives away samples of salami in $\frac{1}{12}$-pound packages. How many pounds of salami will Sal need to make 42 sample packages. Express your answer as an improper fraction *and* as a mixed number.

    Improper fraction:_____

    Mixed number:_____

4. For the company party, Connie ordered 14 pizzas, each divided into 8 slices. At the end of the party, Connie gathered up 18 slices of uneaten pizza. How many pizzas were left over? Express your answer as an improper fraction *and* as a mixed number.

    Improper fraction:_____

    Mixed number:_____

Adding Like Fractions

To add like fractions, add the numerators. Place that sum over the common denominator. If necessary, reduce the answer to lowest terms.

EXAMPLE 1 Add $\frac{13}{16}$ and $\frac{9}{16}$.

STEP 1 The common denominator is 16. Add the numerators. $13 + 9 = 22$

STEP 2 Place the sum over 16.

$$\frac{13}{16} + \frac{9}{16} = \frac{22}{16}$$

STEP 3 Change $\frac{22}{16}$ to a mixed number and reduce the fraction.

$$\frac{22}{16} = 1\frac{6}{16} = 1\frac{3}{8}$$

ANSWER: $1\frac{3}{8}$

EXAMPLE 2 Add $3\frac{5}{8}$ and $2\frac{7}{8}$.

STEP 1 The common denominator is 8. Add the fractions.

STEP 2 Add the whole numbers.

$$\begin{array}{r} 3\frac{5}{8} \\ + 2\frac{7}{8} \\ \hline 5\frac{12}{8} \end{array}$$

STEP 3 Change $\frac{12}{8}$ to a mixed number.

$5\frac{12}{8} = 5 + 1\frac{4}{8}$ $\left(\text{since } \frac{12}{8} = 1\frac{4}{8}\right)$

STEP 4 Add the 1 to the 5. $5 + 1\frac{4}{8} = 6\frac{4}{8}$

Reduce the fraction. $6\frac{4}{8} = 6\frac{1}{2}$

ANSWER: $6\frac{1}{2}$

Add. Reduce each answer to lowest terms.

Fractions Whose Sum Is Less Than 1

1.
$$\begin{array}{r} \frac{1}{3} \\ + \frac{1}{3} \\ \hline \frac{2}{3} \end{array} \qquad \begin{array}{r} \frac{2}{4} \\ + \frac{1}{4} \\ \hline \end{array} \qquad \begin{array}{r} \frac{5}{8} \\ + \frac{2}{8} \\ \hline \end{array} \qquad \begin{array}{r} \frac{2}{5} \\ + \frac{2}{5} \\ \hline \end{array} \qquad \begin{array}{r} \frac{8}{16} \\ + \frac{7}{16} \\ \hline \end{array}$$

2. $\frac{3}{8} + \frac{1}{8} = \frac{4}{8}$ $\qquad \frac{2}{6} + \frac{1}{6} =$ $\qquad \frac{3}{8} + \frac{3}{8} =$ $\qquad \frac{1}{4} + \frac{1}{4} =$ $\qquad \frac{4}{12} + \frac{4}{12} =$

$\frac{4 \div 4}{8 \div 4} = \frac{1}{2}$

3.
$$\begin{array}{r} \frac{3}{8} \\ \frac{2}{8} \\ + \frac{1}{8} \\ \hline \frac{6}{8} \end{array} = \frac{6 \div 2}{8 \div 2} = \frac{3}{4} \qquad \begin{array}{r} \frac{4}{9} \\ \frac{3}{9} \\ + \frac{1}{9} \\ \hline \end{array} \qquad \begin{array}{r} \frac{3}{10} \\ \frac{3}{10} \\ + \frac{2}{10} \\ \hline \end{array} \qquad \begin{array}{r} \frac{2}{6} \\ \frac{1}{6} \\ + \frac{1}{6} \\ \hline \end{array} \qquad \begin{array}{r} \frac{5}{12} \\ \frac{3}{12} \\ + \frac{2}{12} \\ \hline \end{array}$$

Fractions Whose Sum Is a Whole Number

4. $\dfrac{4}{5}$ $\qquad$ $\dfrac{5}{8}$ $\qquad$ $\dfrac{3}{4}$ $\qquad$ $\dfrac{2}{3}$ $\qquad$ $\dfrac{11}{16}$

$+\dfrac{1}{5}$ $\qquad$ $+\dfrac{3}{8}$ $\qquad$ $+\dfrac{1}{4}$ $\qquad$ $+\dfrac{1}{3}$ $\qquad$ $+\dfrac{5}{16}$

$\overline{\dfrac{5}{5}=1}$

5. $\dfrac{3}{4}+\dfrac{2}{4}+\dfrac{3}{4}=\dfrac{8}{4}=2$ $\qquad$ $\dfrac{2}{3}+\dfrac{2}{3}+\dfrac{2}{3}=$ $\qquad$ $\dfrac{7}{8}+\dfrac{6}{8}+\dfrac{5}{8}+\dfrac{6}{8}=$

Fractions Whose Sum Is Greater Than 1

6. $\dfrac{3}{4}$ $\qquad$ $\dfrac{2}{3}$ $\qquad$ $\dfrac{7}{8}$ $\qquad$ $\dfrac{11}{12}$ $\qquad$ $\dfrac{5}{6}$

$+\dfrac{2}{4}$ $\qquad$ $+\dfrac{2}{3}$ $\qquad$ $+\dfrac{6}{8}$ $\qquad$ $+\dfrac{6}{12}$ $\qquad$ $+\dfrac{2}{6}$

$\overline{\dfrac{5}{4}=1\dfrac{1}{4}}$

7. $\dfrac{3}{4}$ $\qquad$ $\dfrac{6}{8}$ $\qquad$ $\dfrac{5}{6}$ $\qquad$ $\dfrac{9}{10}$ $\qquad$ $\dfrac{8}{12}$

$+\dfrac{3}{4}$ $\qquad$ $+\dfrac{4}{8}$ $\qquad$ $+\dfrac{5}{6}$ $\qquad$ $+\dfrac{6}{10}$ $\qquad$ $+\dfrac{6}{12}$

$\overline{\dfrac{6}{4}=1\dfrac{2}{4}=1\dfrac{1}{2}}$

Adding Fractions and Mixed Numbers

8. $1\dfrac{3}{4}$ $\qquad$ $2\dfrac{2}{3}$ $\qquad$ $1\dfrac{3}{8}$ $\qquad$ $5\dfrac{1}{2}$ $\qquad$ $2\dfrac{9}{16}$

$+\dfrac{3}{4}$ $\qquad$ $+\dfrac{2}{3}$ $\qquad$ $+\dfrac{7}{8}$ $\qquad$ $+\dfrac{1}{2}$ $\qquad$ $+\dfrac{11}{16}$

$\overline{1\dfrac{6}{4}=1+1\dfrac{2}{4}}$

$=2\dfrac{2}{4}=2\dfrac{1}{2}$

9. $3\dfrac{4}{6}$ $\qquad$ $2\dfrac{5}{8}$ $\qquad$ $4\dfrac{3}{5}$ $\qquad$ $5\dfrac{11}{12}$ $\qquad$ $8\dfrac{9}{16}$

$+\dfrac{5}{6}$ $\qquad$ $+1\dfrac{7}{8}$ $\qquad$ $+2\dfrac{2}{5}$ $\qquad$ $+3\dfrac{9}{12}$ $\qquad$ $+7\dfrac{5}{16}$

$\overline{5\dfrac{9}{6}=5+1\dfrac{3}{6}}$

$=6\dfrac{3}{6}=6\dfrac{1}{2}$

Subtracting Like Fractions

To subtract like fractions, subtract the numerators. Place that difference over the denominator. If necessary, reduce the answer to lowest terms.

EXAMPLE 1 Subtract $\frac{3}{8}$ from $\frac{7}{8}$.

STEP 1 The common denominator is 8. Subtract the numerators.

STEP 2 Place the difference (4) over the denominator.

$$\begin{array}{r} \frac{7}{8} \\ -\ \frac{3}{8} \\ \hline \frac{4}{8} \end{array}$$

STEP 3 Reduce the fraction. $\frac{4 \div 4}{8 \div 4} = \frac{1}{2}$

ANSWER: $\frac{1}{2}$

EXAMPLE 2 Subtract $2\frac{1}{4}$ from $5\frac{3}{4}$.

STEP 1 The common denominator is 4. Subtract the fractions.

STEP 2 Subtract the whole numbers.

$$\begin{array}{r} 5\frac{3}{4} \\ -\ 2\frac{1}{4} \\ \hline 3\frac{2}{4} \end{array}$$

STEP 3 Reduce the fraction. $\frac{2 \div 2}{4 \div 2} = \frac{1}{2}$

ANSWER: $3\frac{1}{2}$

Subtract. Reduce each answer to lowest terms.

1. $\quad\frac{3}{4}\qquad\qquad\frac{6}{8}\qquad\qquad\frac{4}{5}\qquad\qquad\frac{7}{9}\qquad\qquad\frac{9}{10}$

$\quad-\frac{2}{4}\qquad\quad-\frac{3}{8}\qquad\quad-\frac{2}{5}\qquad\quad-\frac{5}{9}\qquad\quad-\frac{6}{10}$

2. $\frac{6}{8} - \frac{2}{8} = \qquad\quad \frac{8}{9} - \frac{2}{9} = \qquad\quad \frac{11}{12} - \frac{5}{12} = \qquad\quad \frac{13}{16} - \frac{9}{16} =$

Subtracting Fractions from Mixed Numbers

3. $\quad 2\frac{7}{8}\qquad\qquad 1\frac{3}{4}\qquad\qquad 3\frac{2}{3}\qquad\qquad 1\frac{9}{10}\qquad\qquad 2\frac{11}{12}$

$\quad-\frac{3}{8}\qquad\quad-\frac{1}{4}\qquad\quad-\frac{1}{3}\qquad\quad-\frac{5}{10}\qquad\quad-\frac{3}{12}$

Subtracting Mixed Numbers

4. $\quad 3\frac{3}{4}\qquad\qquad 4\frac{7}{8}\qquad\qquad 6\frac{5}{6}\qquad\qquad 12\frac{13}{16}\qquad\qquad 23\frac{11}{12}$

$\quad-2\frac{1}{4}\qquad\quad-2\frac{1}{8}\qquad\quad-3\frac{3}{6}\qquad\quad-4\frac{3}{16}\qquad\quad-15\frac{9}{12}$

Subtracting Fractions from Whole Numbers

To subtract a fraction from a whole number, *regroup* 1 from the whole number. Rewrite the 1 as a fraction, using the same denominator as the fraction you're subtracting. Then subtract as usual.

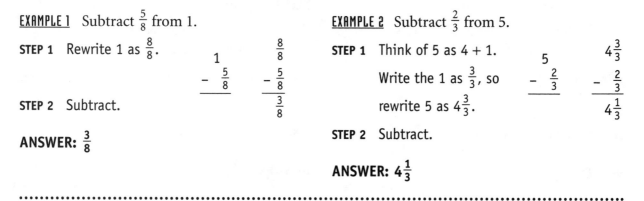

EXAMPLE 1 Subtract $\frac{5}{8}$ from 1.

STEP 1 Rewrite 1 as $\frac{8}{8}$.

$$\begin{array}{r} 1 \\ -\ \frac{5}{8} \\ \hline \end{array} \qquad \begin{array}{r} \frac{8}{8} \\ -\ \frac{5}{8} \\ \hline \frac{3}{8} \end{array}$$

STEP 2 Subtract.

ANSWER: $\frac{3}{8}$

EXAMPLE 2 Subtract $\frac{2}{3}$ from 5.

STEP 1 Think of 5 as $4 + 1$.
Write the 1 as $\frac{3}{3}$, so rewrite 5 as $4\frac{3}{3}$.

$$\begin{array}{r} 5 \\ -\ \frac{2}{3} \\ \hline \end{array} \qquad \begin{array}{r} 4\frac{3}{3} \\ -\ \frac{2}{3} \\ \hline 4\frac{1}{3} \end{array}$$

STEP 2 Subtract.

ANSWER: $4\frac{1}{3}$

Rewrite each whole number by regrouping 1 from the whole number. Use the indicated denominator.

1. $1 = \dfrac{}{3}$ $1 = \dfrac{}{4}$ $1 = \dfrac{}{12}$ $1 = \dfrac{}{2}$ $1 = \dfrac{}{6}$

2. $3 = 2\dfrac{}{8}$ $2 = 1\dfrac{}{3}$ $5 = 4\dfrac{}{6}$ $7 = 6\dfrac{}{2}$ $12 = 11\dfrac{}{4}$

Subtract.

3. $\begin{array}{r} 1 \\ -\ \frac{3}{4} \\ \hline \end{array}$ $\begin{array}{r} 1 \\ -\ \frac{5}{8} \\ \hline \end{array}$ $\begin{array}{r} 1 \\ -\ \frac{7}{16} \\ \hline \end{array}$ $\begin{array}{r} 1 \\ -\ \frac{1}{3} \\ \hline \end{array}$ $\begin{array}{r} 1 \\ -\ \frac{1}{2} \\ \hline \end{array}$

4. $\begin{array}{r} 2 \\ -\ \frac{1}{3} \\ \hline \end{array}$ $\begin{array}{r} 5 \\ -\ \frac{3}{4} \\ \hline \end{array}$ $\begin{array}{r} 3 \\ -\ \frac{5}{8} \\ \hline \end{array}$ $\begin{array}{r} 7 \\ -\ \frac{3}{10} \\ \hline \end{array}$ $\begin{array}{r} 6 \\ -\ \frac{5}{16} \\ \hline \end{array}$

5. $\begin{array}{r} 4 \\ -\ \frac{3}{5} \\ \hline \end{array}$ $\begin{array}{r} 9 \\ -\ \frac{2}{3} \\ \hline \end{array}$ $\begin{array}{r} 8 \\ -\ \frac{7}{8} \\ \hline \end{array}$ $\begin{array}{r} 11 \\ -\ \frac{7}{10} \\ \hline \end{array}$ $\begin{array}{r} 12 \\ -\ \frac{11}{16} \\ \hline \end{array}$

Subtracting Fractions by Regrouping

Sometimes you need to subtract a mixed number from a whole number or from a mixed number with a "too small" fractional part. Before you can subtract, you need to regroup the top whole number.

EXAMPLE   Subtract $2\frac{3}{4}$ from $6\frac{1}{4}$.

STEP 1   Regroup 6 as $5\frac{4}{4}$. Then $6\frac{1}{4} = 5\frac{4}{4} + \frac{1}{4}$.

$$\begin{array}{r} 5\frac{4}{4} + \frac{1}{4} \\ - 2\frac{3}{4} \\ \hline \end{array}$$

STEP 2   Combine the top fractions. $\frac{4}{4} + \frac{1}{4} = \frac{5}{4}$, so $6\frac{1}{4} = 5\frac{5}{4}$.

$$\begin{array}{r} 5\frac{5}{4} \\ - 2\frac{3}{4} \\ \hline \end{array}$$

STEP 3   Subtract the fractions and subtract the whole numbers. Reduce if needed.

$$\begin{array}{r} 5\frac{5}{4} \\ - 2\frac{3}{4} \\ \hline 3\frac{2}{4} = 3\frac{1}{2} \end{array}$$

ANSWER: $3\frac{1}{2}$

Subtract. Reduce each answer to lowest terms.

1. $\begin{array}{r} 3\frac{1}{4} \\ - \frac{3}{4} \\ \hline \end{array}$
$\qquad\qquad$ $\begin{array}{r} 2\frac{1}{3} \\ - \frac{2}{3} \\ \hline \end{array}$
$\qquad\qquad$ $\begin{array}{r} 1\frac{3}{5} \\ - \frac{4}{5} \\ \hline \end{array}$

2. $\begin{array}{r} 5 \\ - 2\frac{1}{2} \\ \hline \end{array}$
$\qquad\qquad$ $\begin{array}{r} 3 \\ - 1\frac{2}{3} \\ \hline \end{array}$
$\qquad\qquad$ $\begin{array}{r} 5 \\ - 3\frac{5}{8} \\ \hline \end{array}$

3. $\begin{array}{r} 4\frac{1}{6} \\ - 3\frac{5}{6} \\ \hline \end{array}$
$\qquad\qquad$ $\begin{array}{r} 8\frac{5}{16} \\ - 4\frac{11}{16} \\ \hline \end{array}$
$\qquad\qquad$ $\begin{array}{r} 7\frac{3}{8} \\ - 2\frac{7}{8} \\ \hline \end{array}$

4. $\begin{array}{r} 14\frac{1}{4} \\ - 8\frac{2}{4} \\ \hline \end{array}$
$\qquad\qquad$ $\begin{array}{r} 19\frac{3}{8} \\ - 10\frac{4}{8} \\ \hline \end{array}$
$\qquad\qquad$ $\begin{array}{r} 26\frac{5}{12} \\ - 18\frac{8}{12} \\ \hline \end{array}$

Adding and Subtracting Unlike Fractions

To add or subtract unlike fractions, rewrite them as like fractions. Then add or subtract as usual. (To review changing unlike fractions to like fractions, review pages 66–67.

EXAMPLE 1 Add $\frac{1}{2}$ and $\frac{2}{3}$.

STEP 1 Comparing multiples, choose 6 as the lowest common denominator. Then rewrite $\frac{1}{2}$ and $\frac{2}{3}$ as *like fractions*.

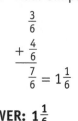

2: 2 4 6
3: 3 6 9

$$\frac{1}{2} = \frac{3}{6}$$
$$+\frac{2}{3} = +\frac{4}{6}$$

STEP 2 Add. Simplify the answer if needed.

$$\frac{3}{6}$$
$$+\frac{4}{6}$$
$$\frac{7}{6} = 1\frac{1}{6}$$

ANSWER: $1\frac{1}{6}$

EXAMPLE 2 Subtract $2\frac{1}{2}$ from $7\frac{3}{8}$.

STEP 1 Write $\frac{1}{2}$ and $\frac{3}{8}$ as *like fractions* with 8 as the lowest common denominator.

$$7\frac{3}{8} = 7\frac{3}{8}$$
$$-2\frac{1}{2} = -2\frac{4}{8}$$

STEP 2 The bottom fraction $\frac{4}{8}$ is greater than the top fraction $\frac{3}{8}$, so regroup 7 as $6\frac{8}{8}$. Then write $6\frac{8}{8} + \frac{3}{8}$ as $6\frac{11}{8}$. Subtract the fractions and the whole numbers.

$$7\frac{3}{8} = 6\frac{8}{8} + \frac{3}{8} = 6\frac{11}{8}$$
$$-2\frac{4}{8} \qquad\qquad = -2\frac{4}{8}$$
$$\qquad\qquad\qquad\qquad 4\frac{7}{8}$$

ANSWER: $4\frac{7}{8}$

Add or subtract as indicated. Simplify your answer if needed.

1. $\frac{1}{2}$ $\frac{2}{3}$ $\frac{3}{4}$ $\frac{7}{8}$ $\frac{7}{8}$
    $+\frac{1}{4}$ $+\frac{1}{6}$ $+\frac{5}{8}$ $+\frac{1}{2}$ $+\frac{11}{16}$

2. $\frac{9}{10}$ $\frac{3}{4}$ $\frac{5}{6}$ $\frac{7}{8}$ $\frac{15}{16}$
    $-\frac{3}{5}$ $-\frac{5}{8}$ $-\frac{1}{3}$ $-\frac{3}{4}$ $-\frac{7}{8}$

3. $\frac{2}{3}$ $\frac{2}{3}$ $\frac{3}{5}$ $\frac{5}{6}$ $\frac{1}{2}$
    $+\frac{1}{4}$ $-\frac{3}{5}$ $+\frac{2}{3}$ $-\frac{3}{4}$ $-\frac{3}{7}$

Add each pair of mixed numbers. As your first step, rewrite unlike fractions as like fractions. Add fractions and whole numbers separately; then simplify.

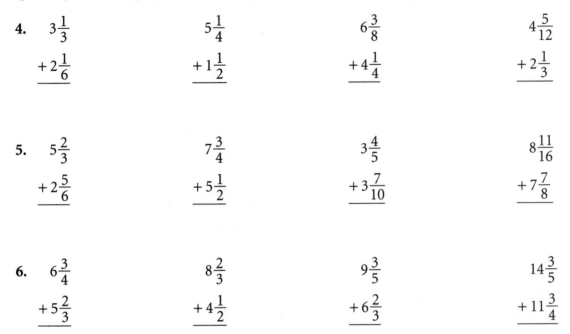

4.　$3\frac{1}{3}$　　　　$5\frac{1}{4}$　　　　$6\frac{3}{8}$　　　　$4\frac{5}{12}$

　　$+\,2\frac{1}{6}$　　　$+\,1\frac{1}{2}$　　　$+\,4\frac{1}{4}$　　　$+\,2\frac{1}{3}$

5.　$5\frac{2}{3}$　　　　$7\frac{3}{4}$　　　　$3\frac{4}{5}$　　　　$8\frac{11}{16}$

　　$+\,2\frac{5}{6}$　　　$+\,5\frac{1}{2}$　　　$+\,3\frac{7}{10}$　　　$+\,7\frac{7}{8}$

6.　$6\frac{3}{4}$　　　　$8\frac{2}{3}$　　　　$9\frac{3}{5}$　　　　$14\frac{3}{5}$

　　$+\,5\frac{2}{3}$　　　$+\,4\frac{1}{2}$　　　$+\,6\frac{2}{3}$　　　$+\,11\frac{3}{4}$

Subtract each pair of mixed numbers. As your first step, rewrite unlike fractions as like fractions. Regroup if needed; then subtract the fractions and subtract the whole numbers.

7.　$9\frac{7}{8}$　　　　$5\frac{2}{3}$　　　　$4\frac{11}{12}$　　　　$13\frac{3}{4}$

　　$-\,5\frac{3}{4}$　　　$-\,2\frac{1}{6}$　　　$-\,1\frac{1}{4}$　　　$-\,9\frac{1}{2}$

8.　$8\frac{3}{8}$　　　　$3\frac{1}{3}$　　　　$7\frac{5}{12}$　　　　$15\frac{1}{4}$

　　$-\,5\frac{3}{4}$　　　$-\,1\frac{5}{6}$　　　$-\,2\frac{3}{4}$　　　$-\,10\frac{1}{2}$

9.　$9\frac{1}{4}$　　　　$6\frac{1}{3}$　　　　$11\frac{1}{5}$　　　　$23\frac{1}{3}$

　　$-\,4\frac{2}{3}$　　　$-\,5\frac{1}{2}$　　　$-\,3\frac{1}{2}$　　　$-\,14\frac{4}{5}$

Applying Your Skills

Solve the following word problems. They involve adding and subtracting fractions and mixed numbers.

1. Allie ran $2\frac{3}{4}$ miles on Monday, $3\frac{1}{4}$ miles on Wednesday, and $4\frac{3}{4}$ miles on Friday. How many total miles did Allie run this week?

2. Two wooden beams sit side by side. One is $12\frac{15}{16}$ inches wide, and the other is $8\frac{7}{8}$ inches wide. What is their combined width?

3. The clerk at Valley Fabric cut $4\frac{1}{3}$ yards of cloth from a bolt of fabric containing $9\frac{2}{3}$ yards. After she removed this piece, how much fabric was left on the bolt?

4. Connie ordered carpet that is $\frac{3}{4}$ inch thick. The accompanying pad is $\frac{5}{8}$ inch thick. How thick are the carpet and pad together?

5. Joe's pickup is able to carry $\frac{7}{8}$ ton of rocks. Frieda's pickup can carry $\frac{3}{4}$ ton of rocks. How much more rock can Joe's pickup haul in each load than Frieda's?

6. At Twin Pines Lumber, Katrina bought $2\frac{1}{4}$ pounds of #6 nails and $1\frac{2}{3}$ pounds of #8 nails. What total weight of nails did she buy?

7. Ed had planned to buy $8\frac{1}{2}$ pounds of apples for the picnic. Instead, he bought a bargain bag that weighed $6\frac{2}{3}$ pounds. How many fewer pounds of apples did Ed buy than he had planned?

8. During the first three weeks in November, Eugene recorded the following amounts of rainfall: week 1, $2\frac{3}{4}$ inches; week 2, $1\frac{2}{3}$ inches; week 3, $2\frac{7}{12}$ inches. How much rain fell during these three weeks?

9. A biscuit recipe calls for $2\frac{1}{4}$ cups of flour. If Lauren has only $\frac{7}{8}$ cup of flour, how much more does she need?

10. Meg's new curtain rod extends to a width of $51\frac{5}{8}$ inches. How much wider is the rod than Meg's $46\frac{13}{16}$-inch-wide bedroom window?

11. Following a recipe, Armondo mixed $2\frac{1}{2}$ cups of flour together with $\frac{3}{4}$ cup of sugar and $\frac{1}{3}$ cup of butter. He combined these ingredients in a bowl that holds 8 cups. How many more cups of ingredients can this bowl hold before it is full?

12. The newspaper said that 400 of the town's high school students already had chicken pox. Of this group, $\frac{3}{5}$ contracted the virus before going into the 4th grade, and $\frac{1}{3}$ got it while in the 4th, 5th, or 6th grade. What fraction of this group of students got chicken pox after the 6th grade?

13. For presents, Alexis makes serving trays.

    a. What is the length and width of the tray Alexis makes?

    length: _____   width: _____

    b. How much longer is the length of the tray than its width?

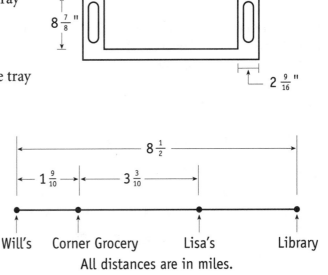

14. The map shows that Corner Grocery is between Will's house and Lisa's house.

    a. How far does Lisa live from Will?

    b. How far is the library from Corner Grocery?

15. Find the total overtime hours worked on a 3-day weekend by each employee.

Workday	J. Blakely	R. Craig
Saturday	$1\frac{1}{3}$ hr	$\frac{3}{4}$ hr
Sunday	$\frac{1}{2}$ hr	$1\frac{1}{2}$ hr
Monday	$2\frac{5}{6}$ hr	$\frac{2}{3}$ hr

16. On the first 3 weeks of her diet, Wanda lost weight as follows: week 1, $2\frac{1}{2}$ lb; week 2, $1\frac{1}{4}$ lb; and week 3, $\frac{7}{8}$ lb. Circle the letter of the equation below that tells Wanda's average weekly weight loss.

    a. $y = (\frac{5}{2} + \frac{5}{4} + \frac{7}{8}) \times 3$ c. $y = (\frac{5}{2} + \frac{5}{4} + \frac{7}{8}) \div 2$

    b. $y = (\frac{5}{2} + \frac{5}{4} + \frac{7}{8}) \div \frac{1}{3}$ d. $y = (\frac{5}{2} + \frac{5}{4} + \frac{7}{8}) \div 3$

Problem Solver: Identifying Missing Information

Some problems on a math test do not give all the necessary information you need to find a solution. Here are two ways these problems are written.

Not Enough Information Given

EXAMPLE 1 City Center Hardware advertised, "All nails, $1.78 per pound." Francesca bought $7\frac{3}{4}$ pounds of #16 nails, $4\frac{1}{2}$ pounds of #12 nails, and a full box of #8 nails. How many pounds of nails did Francesca purchase?

a. $11\frac{1}{4}$ c. $12\frac{1}{4}$ e. not enough information given

b. $11\frac{4}{6}$ d. $12\frac{4}{6}$

The correct answer choice is **e.** To solve this problem, you must know the weight of a full box of #8 nails. The problem does not tell you this weight.

Before choosing **not enough information given,** be careful! Be sure you can identify what information is missing. Only by doing this can you be sure this choice is correct.

What More Do You Need to Know?

EXAMPLE 2 City Center Hardware advertised, "All nails, $1.78 per pound." Francesca bought $7\frac{3}{4}$ pounds of #16 nails, $4\frac{1}{2}$ pounds of #12 nails, and a full box of #8 nails. What more do you need to know to find out how many pounds of nails Francesca purchased?

a. the number of #8 nails Francesca bought   d. the weight of a full box of #16 nails
b. the regular price of #8 nails   e. the weight of a full box of #12 nails
c. the weight of a full box of #8 nails

The correct answer choice is **c.** As in Example 1, you need to know the weight of a full box of #8 nails before you can determine how many pounds of nails Francesca purchased.

Comparing Examples 1 and 2

Similarities: Both give the same information. Both are concerned with finding how many pounds of nails Francesca purchased.

Differences: Example 1 checks your ability to recognize that the given information is incomplete. Example 2 asks you to identify the particular missing information from a list of choices.

In each problem below, identify the missing information.

1. As part of his diet, Eldon counts the calories of foods he eats. How many calories will Eldon consume in a lunch consisting of a 485-calorie hamburger, a 175-calorie soft drink, and french fries?

   Missing information:

2. Twenty percent of the students in Mrs. Jones's preschool class are 3 years old, thirty-five percent are 4 years old, and the rest are either 5 or 6 years old. What total percent of Mrs. Jones's class is made up of 4- *and* 5-year-olds?

   Missing information:

3. One-fourth of Dave's salary goes for rent, one-third for food, and one-tenth for a car payment. What total fraction of his salary does Dave spend for rent, food, car payment, and utilities?

   Missing information:

4. When the price of cat food was reduced from $4.49 per pound, Shauna bought 4 pounds. She also bought 3 pounds of cat litter for $1.39 per pound. What total amount did Shauna pay for these two items?

   Missing information:

5. To help pay for a new pickup, Lewis borrowed $8,000 from his credit union. He must pay back the loan in monthly payments of $208.75. To repay the loan, how much will Lewis actually pay?

   Missing information:

6. George bought $6\frac{3}{4}$ pounds of apples, a large bag of pears, and $7\frac{1}{2}$ pounds of peaches. How many more pounds of peaches did George buy than pears?

   Missing information:

7. Each Monday, Wednesday, and Friday evening, Lucas attends an adult math class. On these days, he studies for 2 hours each day. Each Sunday, Tuesday, and Thursday, Lucas spends additional time studying to prepare for the next day's class. How many hours does Lucas study math each week?

   Missing information:

Multiplying Fractions

To multiply a number by a fraction is to find a part of that number.

As shown at the right, $\frac{1}{2}$ of $\frac{1}{4}$ is $\frac{1}{8}$.

Multiplying fractions is the same for both like fractions and unlike fractions.

- Multiply the numerators to find the numerator of the answer.
- Multiply the denominators to find the denominator of the answer.

$\frac{1}{2}$ of $\frac{1}{4}$ is $\frac{1}{8}$.

EXAMPLE 1 $\frac{3}{4} \times \frac{1}{2} =$

    STEP 1 Multiply the numerators.

$$\frac{3}{4} \times \frac{1}{2} = \frac{3}{}$$

    STEP 2 Multiply the denominators.

$$\frac{3}{4} \times \frac{1}{2} = \frac{3}{8}$$

ANSWER: $\frac{3}{8}$

EXAMPLE 2 $\frac{2}{3} \times \frac{3}{8} =$

    STEP 1 Multiply the numerators.

$$\frac{2}{3} \times \frac{3}{8} = \frac{6}{}$$

    STEP 2 Multiply the denominators.

$$\frac{2}{3} \times \frac{3}{8} = \frac{6}{24}$$

    STEP 3 Reduce the answer.

$$\frac{6}{24} = \frac{6 \div 6}{24 \div 6} = \frac{1}{4}$$

ANSWER: $\frac{1}{4}$

Multiply. Reduce your answers.

1. $\frac{1}{2} \times \frac{2}{3} =$ $\frac{3}{4} \times \frac{1}{3} =$ $\frac{2}{3} \times \frac{7}{10} =$

2. $\frac{3}{5} \times \frac{1}{9} =$ $\frac{3}{8} \times \frac{1}{2} =$ $\frac{2}{5} \times \frac{7}{8} =$

3. $\frac{3}{4} \times \frac{2}{3} =$ $\frac{3}{7} \times \frac{3}{10} =$ $\frac{3}{4} \times \frac{5}{6} =$

4. $\frac{3}{4} \times \frac{3}{4} =$ $\frac{2}{4} \times \frac{3}{10} =$ $\frac{11}{12} \times \frac{3}{4} =$

5. $\frac{1}{3} \times \frac{3}{4} \times \frac{2}{5} =$ $\frac{2}{3} \times \frac{1}{2} \times \frac{3}{4} =$ $\frac{3}{5} \times \frac{2}{3} \times \frac{1}{4} =$

Canceling to Simplify Multiplication

When multiplying fractions, you can often use a shortcut called **canceling.** To cancel, divide both the numerator and the denominator by the same number. The numerator and denominator can be in the same fraction or they can be parts of two different fractions. Canceling is similar to reducing a fraction.

<u>EXAMPLE 1</u> $\frac{5}{6} \times \frac{3}{4} =$

STEP 1 Ask, "Can a numerator and a denominator be divided by the same number?" Yes. 6 and 3 can be divided by 3.

STEP 2 Divide. $6 \div 3 = 2$ and $3 \div 3 = 1$ After you cancel, multiply the new fractions.

$$\frac{5}{\cancel{6}_2} \times \frac{\cancel{3}^1}{4} = \frac{5}{2} \times \frac{1}{4} = \frac{5}{8}$$

ANSWER: $\frac{5}{8}$

<u>EXAMPLE 2</u> $\frac{4}{9} \times \frac{3}{8} =$

STEP 1 Divide both the 4 and the 8 by 4.

STEP 2 Divide both the 9 and the 3 by 3.

$$\frac{{}^1\cancel{4}}{{}_3\cancel{9}} \times \frac{{}^1\cancel{3}}{\cancel{8}_2} = \frac{1}{3} \times \frac{1}{2}$$

STEP 3 Multiply the new fractions.

$$\frac{1}{3} \times \frac{1}{2} = \frac{1}{6}$$

ANSWER: $\frac{1}{6}$

....

Multiply. Use canceling as your first step.

6. $\frac{5}{6} \times \frac{3}{4} =$ $\frac{2}{5} \times \frac{11}{12} =$ $\frac{7}{8} \times \frac{6}{7} =$

7. $\frac{3}{4} \times \frac{8}{9} =$ $\frac{1}{4} \times \frac{2}{5} =$ $\frac{5}{9} \times \frac{3}{10} =$

8. $\frac{5}{12} \times \frac{7}{15} =$ $\frac{3}{4} \times \frac{4}{15} =$ $\frac{7}{8} \times \frac{5}{7} =$

9. $\frac{3}{8} \times \frac{4}{6} =$ $\frac{1}{2} \times \frac{3}{4} \times \frac{8}{9} =$ $\frac{3}{4} \times \frac{4}{5} \times \frac{10}{12} =$

Multiplying Fractions, Whole Numbers, and Mixed Numbers

When you multiply any combination of fractions, whole numbers, and mixed numbers, the first step is to rewrite every whole number or mixed number as an improper fraction.

EXAMPLE 1 $\frac{2}{3} \times 5 =$

STEP 1 Rewrite the whole number 5 as a fraction. Write 5 as $\frac{5}{1}$.

STEP 2 Multiply the fractions.

$$\frac{2}{3} \times \frac{5}{1} = \frac{10}{3} \leftarrow \frac{\text{numerator times numerator}}{\text{denominator times denominator}}$$

ANSWER: $\frac{10}{3} = 3\frac{1}{3}$

EXAMPLE 2 $3\frac{1}{2} \times \frac{1}{4} =$

STEP 1 Write $3\frac{1}{2}$ as an improper fraction.

a. Change 3 to an improper fraction with a denominator of 2.

$$3 = \frac{6}{2}$$

b. Add $\frac{6}{2}$ and $\frac{1}{2}$.

$$3\frac{1}{2} = \frac{6}{2} + \frac{1}{2} = \frac{7}{2}$$

STEP 2 Multiply $\frac{7}{2}$ by $\frac{1}{4}$.

$$\frac{7}{2} \times \frac{1}{4} = \frac{7}{8}$$

ANSWER: $\frac{7}{8}$

Write each whole number as an improper fraction.

1. $2 =$ $\qquad$ $4 =$ $\qquad$ $7 =$ $\qquad$ $5 =$ $\qquad$ $3 =$ $\qquad$ $12 =$

Write each mixed number as an improper fraction.

2. $1\frac{2}{3} = \frac{3}{3} + \frac{2}{3} = \frac{5}{3}$ $\quad$ $3\frac{1}{4} =$ $\qquad$ $5\frac{3}{8} =$ $\qquad$ $4\frac{3}{5} =$ $\qquad$ $8\frac{1}{3} =$ $\qquad$ $11\frac{1}{2} =$

Multiply. Use cancellation when possible. Simplify your answers.

3. $\frac{1}{2} \times 4 = \frac{1}{\underset{1}{2}} \times \frac{\overset{2}{4}}{1} = \frac{2}{1} = 2$ $\qquad$ $\frac{2}{3} \times 5 =$ $\qquad$ $6 \times \frac{2}{5} =$

4. $\frac{3}{4} \times 8 =$ $\qquad$ $\frac{1}{3} \times 9 =$ $\qquad$ $10 \times \frac{4}{5} =$

5. $12 \times \frac{2}{3} =$ $\qquad$ $20 \times \frac{4}{5} =$ $\qquad$ $\frac{5}{6} \times 18 =$

Math Tip

Some students like to use this shortcut for changing a mixed number to an improper fraction:

Numerator: Multiply the whole number by the denominator and add the numerator.

$$2\frac{3}{4} = \frac{(4 \times 2) + 3}{4} = \frac{11}{4} \qquad 5\frac{2}{3} = \frac{(3 \times 5) + 2}{3} = \frac{17}{3}$$

Denominator: Remains the same.

Use the shortcut to change each mixed number to an improper fraction.

6. $5\frac{1}{2} =$ $\qquad$ $2\frac{7}{8} =$ $\qquad$ $4\frac{3}{5} =$

Multiply. Use cancellation when possible. Simplify your answers.

7. $2\frac{2}{3} \times 4 =$ $\qquad$ $1\frac{3}{4} \times 3 =$ $\qquad$ $4\frac{1}{2} \times 6 =$

8. $6 \times 3\frac{1}{4} =$ $\qquad$ $4\frac{5}{12} \times 3 =$ $\qquad$ $6\frac{2}{5} \times 10 =$

9. $2\frac{1}{2} \times \frac{2}{3} =$ $\qquad$ $1\frac{5}{8} \times \frac{4}{5} =$ $\qquad$ $\frac{3}{4} \times 3\frac{1}{8} =$

10. $\frac{5}{8} \times 3\frac{1}{4} =$ $\qquad$ $4\frac{1}{2} \times \frac{5}{6} =$ $\qquad$ $\frac{4}{5} \times 1\frac{11}{12} =$

11. $3\frac{1}{2} \times 2\frac{2}{3} =$ $\qquad$ $4\frac{1}{4} \times 1\frac{3}{4} =$ $\qquad$ $5\frac{1}{2} \times 3\frac{5}{8} =$

12. $4\frac{1}{3} \times 3\frac{7}{8} =$ $\qquad$ $1\frac{15}{16} \times 2\frac{1}{4} =$ $\qquad$ $2\frac{3}{4} \times 5\frac{1}{4} =$

Dividing Fractions

Dividing is finding out how many times one amount can be found in a second amount. This is also true with fractions.

To find out how many $\frac{1}{4}$-inch-long sections are in a length of $1\frac{1}{2}$ inches, divide $1\frac{1}{2}$ by $\frac{1}{4}$.

As shown at the right, there are six $\frac{1}{4}$s in $1\frac{1}{2}$.

Dividing fractions involves just one more step than multiplying fractions.

To divide by a fraction, invert the divisor (the number you're dividing by) and change the division sign to a multiplication sign. Then multiply.

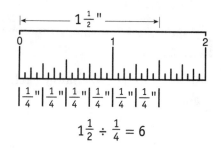

$$1\frac{1}{2} \div \frac{1}{4} = 6$$

Inverting the Divisor

To **invert** means to turn a fraction upside down. When you invert a fraction, you switch the top and bottom numbers.

Inverting $\frac{2}{3}$

$$\frac{2}{3} \quad \frac{3}{2}$$

Before inverting a divisor, be sure it is written as a proper or improper fraction.

- Change a whole-number divisor to a fraction by placing the whole number over 1.

- Change a mixed-number divisor to an improper fraction.

Type of Divisor	Example	Write the Divisor as a Fraction	Invert the Divisor and Change the Sign
proper fraction	$\frac{3}{4} \div \frac{7}{8}$	$\frac{3}{4} \div \frac{7}{8}$	$\frac{3}{4} \times \frac{8}{7}$
improper fraction	$\frac{3}{5} \div \frac{3}{2}$	$\frac{3}{5} \div \frac{3}{2}$	$\frac{3}{5} \times \frac{2}{3}$
whole number	$\frac{3}{8} \div 5$	$\frac{3}{8} \div \frac{5}{1}$	$\frac{3}{8} \times \frac{1}{5}$
mixed number	$\frac{15}{16} \div 2\frac{3}{8}$	$\frac{15}{16} \div \frac{19}{8}$	$\frac{15}{16} \times \frac{8}{19}$

Invert each number below.

1. $\frac{2}{3}$ $\qquad$ $\frac{5}{8}$ $\qquad$ $\frac{5}{6}$ $\qquad$ $\frac{1}{2}$ $\qquad$ $\frac{3}{4}$

2. $\frac{4}{3}$ $\qquad$ $\frac{7}{5}$ $\qquad$ $\frac{9}{4}$ $\qquad$ $\frac{5}{2}$ $\qquad$ $\frac{21}{16}$

3. 2 $\qquad$ 3 $\qquad$ 5 $\qquad$ 4 $\qquad$ 7

4. $2\frac{2}{3}$ $\qquad$ $3\frac{1}{4}$ $\qquad$ $4\frac{1}{3}$ $\qquad$ $1\frac{5}{8}$ $\qquad$ $2\frac{3}{16}$

Dividing a Fraction by a Fraction

EXAMPLE $\frac{3}{4} \div \frac{2}{3} =$

STEP 1 Invert the divisor. (The divisor is the second fraction. It is after the division sign.)

STEP 2 Change the division sign to a multiplication sign. Then multiply. Simplify the answer if necessary.

Invert the divisor.

Divide: $\frac{3}{4} \div \frac{2}{3} =$

Change the sign.

$\frac{3}{4} \times \frac{3}{2} = \frac{9}{8}$

ANSWER: $\frac{9}{8} = 1\frac{1}{8}$

· ·

Divide. Simplify your answers.

5. $\frac{1}{2} \div \frac{3}{4} =$ $\qquad$ $\frac{3}{4} \div \frac{1}{3} =$ $\qquad$ $\frac{1}{3} \div \frac{3}{8} =$ $\qquad$ $\frac{2}{3} \div \frac{2}{4} =$

6. $\frac{7}{8} \div \frac{7}{8} =$ $\qquad$ $\frac{3}{6} \div \frac{1}{3} =$ $\qquad$ $\frac{11}{16} \div \frac{1}{4} =$ $\qquad$ $\frac{7}{8} \div \frac{1}{8} =$

7. $\frac{2}{3} \div \frac{1}{2} =$ $\qquad$ $\frac{7}{8} \div \frac{1}{2} =$ $\qquad$ $\frac{4}{5} \div \frac{4}{5} =$ $\qquad$ $\frac{5}{16} \div \frac{1}{16} =$

8. $\frac{12}{8} \div \frac{2}{3} =$ $\qquad$ $\frac{14}{6} \div \frac{1}{2} =$ $\qquad$ $\frac{3}{4} \div \frac{5}{4} =$ $\qquad$ $\frac{4}{3} \div \frac{3}{2} =$

Dividing Fractions, Whole Numbers, and Mixed Numbers

When you are dividing any combination of fractions, mixed numbers, and whole numbers, the first step is to rewrite every mixed number or whole number as an improper fraction.

Dividing with Fractions and Whole Numbers

When you are dividing fractions with whole numbers,

- write the whole number as a fraction with a denominator of 1
- invert the divisor; then multiply the fractions

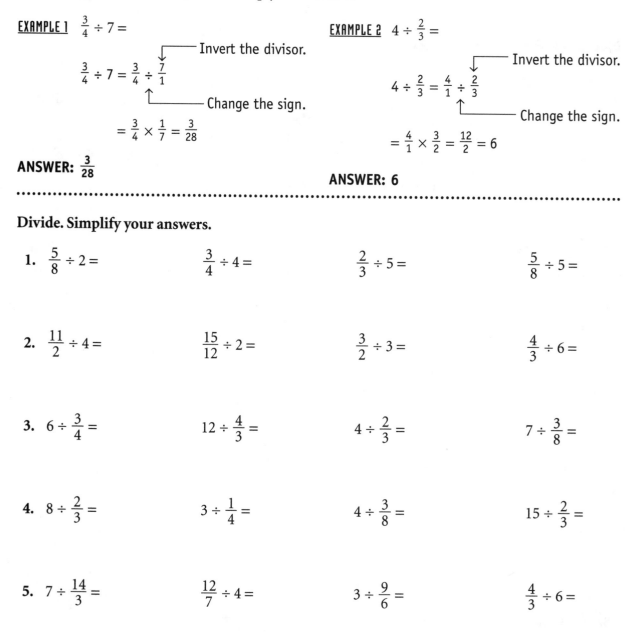

EXAMPLE 1 $\frac{3}{4} \div 7 =$

$\frac{3}{4} \div 7 = \frac{3}{4} \div \frac{7}{1}$ — Invert the divisor.
— Change the sign.

$= \frac{3}{4} \times \frac{1}{7} = \frac{3}{28}$

ANSWER: $\frac{3}{28}$

EXAMPLE 2 $4 \div \frac{2}{3} =$

$4 \div \frac{2}{3} = \frac{4}{1} \div \frac{2}{3}$ — Invert the divisor.
— Change the sign.

$= \frac{4}{1} \times \frac{3}{2} = \frac{12}{2} = 6$

ANSWER: 6

Divide. Simplify your answers.

1. $\frac{5}{8} \div 2 =$ $\frac{3}{4} \div 4 =$ $\frac{2}{3} \div 5 =$ $\frac{5}{8} \div 5 =$

2. $\frac{11}{2} \div 4 =$ $\frac{15}{12} \div 2 =$ $\frac{3}{2} \div 3 =$ $\frac{4}{3} \div 6 =$

3. $6 \div \frac{3}{4} =$ $12 \div \frac{4}{3} =$ $4 \div \frac{2}{3} =$ $7 \div \frac{3}{8} =$

4. $8 \div \frac{2}{3} =$ $3 \div \frac{1}{4} =$ $4 \div \frac{3}{8} =$ $15 \div \frac{2}{3} =$

5. $7 \div \frac{14}{3} =$ $\frac{12}{7} \div 4 =$ $3 \div \frac{9}{6} =$ $\frac{4}{3} \div 6 =$

Dividing with Mixed Numbers

When you are dividing with mixed numbers,

- change each mixed number to an improper fraction
- invert the divisor; then multiply the fractions

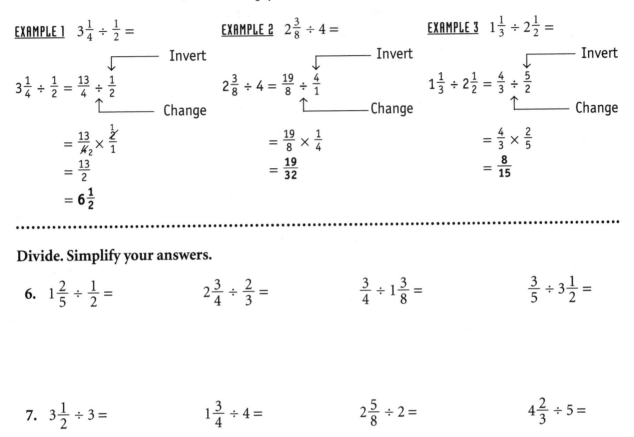

EXAMPLE 1 $3\frac{1}{4} \div \frac{1}{2} =$

$3\frac{1}{4} \div \frac{1}{2} = \frac{13}{4} \div \frac{1}{2}$ — Invert / Change

$= \frac{13}{\cancel{4}_2} \times \frac{\cancel{2}^1}{1}$

$= \frac{13}{2}$

$= 6\frac{1}{2}$

EXAMPLE 2 $2\frac{3}{8} \div 4 =$

$2\frac{3}{8} \div 4 = \frac{19}{8} \div \frac{4}{1}$ — Invert / Change

$= \frac{19}{8} \times \frac{1}{4}$

$= \frac{19}{32}$

EXAMPLE 3 $1\frac{1}{3} \div 2\frac{1}{2} =$

$1\frac{1}{3} \div 2\frac{1}{2} = \frac{4}{3} \div \frac{5}{2}$ — Invert / Change

$= \frac{4}{3} \times \frac{2}{5}$

$= \frac{8}{15}$

....

Divide. Simplify your answers.

6. $1\frac{2}{5} \div \frac{1}{2} =$ $2\frac{3}{4} \div \frac{2}{3} =$ $\frac{3}{4} \div 1\frac{3}{8} =$ $\frac{3}{5} \div 3\frac{1}{2} =$

7. $3\frac{1}{2} \div 3 =$ $1\frac{3}{4} \div 4 =$ $2\frac{5}{8} \div 2 =$ $4\frac{2}{3} \div 5 =$

8. $2\frac{1}{2} \div 1\frac{1}{2} =$ $4\frac{1}{4} \div 8\frac{1}{2} =$ $2\frac{1}{4} \div 3\frac{3}{8} =$ $6\frac{3}{4} \div 2\frac{3}{8} =$

9. $\frac{7}{8} \div 2 =$ $2\frac{3}{4} \div 2 =$ $3 \div 4\frac{1}{2} =$ $4\frac{2}{3} \div 2\frac{1}{12} =$

Applying Your Skills

Solve the following word problems. They involve multiplying and dividing fractions and mixed numbers.

1. A paving crew has $\frac{7}{10}$ mile of road to pave by May 14. The crew hopes to complete $\frac{2}{3}$ of this by May 6. What length of road does the crew hope to finish by May 6?

2. Nora plans to use washers as spacers to raise the corner of an uneven table by $\frac{5}{8}$ inch. How many $\frac{1}{16}$-inch-thick washers will Nora need to level the table?

3. A cookbook recommends $\frac{1}{3}$ hour of cooking time per pound to cook a roast medium well. Following these instructions, how long should a roast weighing $8\frac{1}{2}$ pounds be cooked if the chef wants it medium well?

4. To make curtains for their home, the Brechts bought a $17\frac{1}{2}$-yard bolt of fabric. If it takes $3\frac{1}{4}$ yards of fabric for each window curtain, for how many windows do the Brechts have enough fabric? (**Hint:** Your answer must be a whole number!)

5. Trish, a housepainter, painted $6\frac{1}{2}$ wall panels in one hour. At this rate, how many *complete* wall panels can she paint in $7\frac{3}{4}$ hours?

6. On Tuesday, stock values rose when the Federal Reserve announced a lowering of interest rates. Three stocks whose values changed are shown at the right.

    (**Note:** Each increase stands for the fraction of a dollar that a stock's value increased.)

    What is the *average* increase in value of the three stocks listed?

Stock	Change in Value
ARI	$+\frac{3}{4}$
UMB	$+\frac{1}{2}$
RWM	$+\frac{7}{8}$

7. Spencer built a picnic table, the end view of which is shown. How wide is the table? (**Hint:** Width = 6 table boards + 5 gaps)

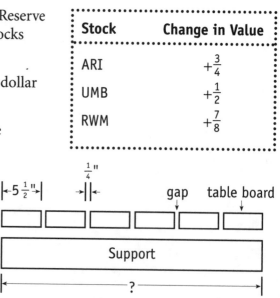

8. Ching-Yan bought $1\frac{1}{4}$ pounds of cheese on sale for $3 per pound and a $6\frac{1}{2}$-pound package of sirloin steaks on sale for $4 per pound. Which expression shows how much Ching-Yan will pay for these items.

    a. $(\frac{5}{4} + \$3) + (\frac{13}{2} + \$4)$ c. $(\frac{5}{4} + \frac{13}{2})$ ($\$3 + \$4$)   e. $(\frac{5}{4})$ ($\$3$) $+ (\frac{13}{2})$ ($\$4$)

    b. $(\frac{5}{4} + \frac{13}{2}) + (\$3 + \$4)$ d. $(\frac{5}{4})$ ($\$4$) $+ (\frac{13}{2})$ ($\$3$)

9. What fraction of their monthly income do the Nelsons spend on each of the following? Reduce each answer to lowest terms.

    Rent: _____   Food: _____   Auto: _____

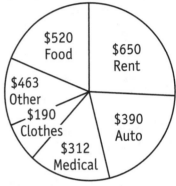

NELSON FAMILY BUDGET
(monthly income: $2,600)

$520 Food
$650 Rent
$463 Other
$190 Clothes
$312 Medical
$390 Auto

10. If the Nelsons donate $\frac{1}{25}$ of their income to charity, how much do they donate each month?

11. For a barbecue, Mike bought $4\frac{1}{2}$ pounds of hamburger. If he divides this into patties weighing about $\frac{1}{4}$ pound, how many patties can he make?

12. Wallpaper comes in rolls that are $1\frac{3}{4}$ feet wide. How many side-by-side strips of wallpaper will Ray need to paper the wall of his den, which is 15 feet wide?

13. Jackie, a jeweler, uses various silver alloys to make earrings, bracelets, and buckles. Fill in the blank lines on the table to show how many complete items of each type she can make with her remaining alloy supply.

Item	Alloy Type	Alloy Supply in Ounces (oz) (a)	Alloy Needed per Item (oz) (b)	Number of Items That Can Be Made (a ÷ b)
Earrings	90/10	7	$\frac{3}{4}$	_____
Bracelets	80/20	$15\frac{1}{2}$	3	_____
Buckles	70/30	$22\frac{1}{2}$	$3\frac{2}{3}$	_____

Relating Fractions and Ratios

A **ratio** is a comparison of two numbers. For example, if there are 8 women and 5 men in your math class, the ratio of women to men is 8 to 5.

You can write the ratio 8 to 5 using symbols in two ways:

- As a fraction, write 8 to 5 as $\frac{8}{5}$.

- With a colon, write 8 to 5 as 8:5.

In words, always read a ratio with the word *to*.
Read the ratios $\frac{8}{5}$ and 8:5 as "eight to five."

To simplify a ratio, write it as a fraction. Then follow these four rules.

1. Reduce the ratio to lowest terms.

    $$6 \text{ to } 10 = \frac{6}{10} = \frac{3}{5}$$

    ANSWER: $\frac{3}{5}$

2. Write a whole-number ratio with a denominator of 1.

    $$4 \text{ to } 1 = \frac{4}{1} \text{ (Leave as is.)}$$

    ANSWER: $\frac{4}{1}$

3. **Do not** write an improper-fraction ratio as a mixed number.

    $$12 \text{ to } 8 = \frac{12}{8} = \frac{3}{2}$$

    ANSWER: $\frac{3}{2}$ (Leave as an improper fraction.)

4. Rewrite a complex ratio as an equal fraction that has only whole numbers.

    $$5\frac{1}{2} \text{ to } 3 = \frac{5\frac{1}{2}}{3}$$

    To rewrite this ratio, divide the top mixed number ($5\frac{1}{2}$) by the bottom number (3).

    $$5\frac{1}{2} \div 3 = \frac{11}{2} \div \frac{3}{1} = \frac{11}{2} \times \frac{1}{3} = \frac{11}{6}$$

    ANSWER: $\frac{11}{6}$

Write each ratio as a fraction in lowest terms.

1. 4 to 6　　　　6 to 8　　　　15 to 3　　　　6 to 4　　　　8 to 16

2. 6 to 2　　　　12 to 8　　　　$4\frac{1}{2}$ to 3　　　　$3\frac{2}{3}$ to 2　　　　$6\frac{1}{2}$ to $4\frac{1}{3}$

To solve a ratio word problem, write the ratio in the same order as it appears in the question.

One-Step Ratio Problem

EXAMPLE 1 Wanda's part-time job pays her $1,200 per month. If she pays $300 each month in rent, what is the ratio of her *rent to* her *income?*

Write the ratio of *rent to income.* Reduce.

$$\frac{\text{rent}}{\text{income}} = \frac{\$300}{\$1,200}$$
$$= \frac{1}{4}$$

ANSWER: $\frac{1}{4}$ *or* **1:4** *or* **1 to 4**

(**Note:** If the question had asked for the ratio of *income to rent,* the answer would be 4 to 1.)

Two-Step Ratio Problem

EXAMPLE 2 On a test of 40 questions, Thurman answered all of the questions and got 6 wrong. What is the ratio of *correct* answers to *incorrect* answers?

STEP 1 Determine how many questions Thurman answered correctly: 40 − 6 = 34 correct.

STEP 2 Write the ratio of *correct to incorrect.*

$$\frac{\text{correct answers}}{\text{incorrect answers}} = \frac{34}{6} = \frac{17}{3}$$

ANSWER: $\frac{17}{3}$ *or* **17:3** *or* **17 to 3**

Solve the following ratio problems.

3. A new compact car gets 48 miles to the gallon during highway driving and 32 miles to the gallon during city driving.

    a. What is the ratio of this car's city mileage to highway mileage?
    b. What is the ratio of this car's highway mileage to city mileage?

4. In a class with 30 students, there are 16 women and 14 men.

    a. What is the ratio of men to women in this class?
    b. What is the ratio of women to men?
    c. What is the ratio of women to the total number of students?

5. Last season the Bulldog basketball team won only 8 of its 22 games. It lost the rest.

    a. What is the ratio of the number of games the team won to the number of games it played?
    b. What is the ratio of the number of games the team won to the number of games it lost?

6. One yard is equal to 3 feet (36 inches). To find a ratio of lengths, both lengths must be expressed in the same unit.

    a. What is the ratio of one yard to one foot? (**Hint:** As your first step, write 1 yard as a number of feet.)
    b. What is the ratio of 8 inches to 2 yards? (**Hint:** As your first step, write 2 yards as a number of inches.)

Problem Solver: Recognizing and Comparing Patterns

Recognizing Patterns

Many types of sequence problems can be solved by looking for patterns. First, determine the common link that relates numbers in the pattern. Then, use that link to continue a **number series** or to predict a future value based on a **trend** in data.

Number Series

EXAMPLE 1 What is the next number in the following number series?

6, 12, 24, . . .

STEP 1 Find the common link.

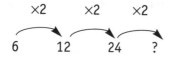

Link: Multiply by 2.

STEP 2 Find the next number.

Multiply 24 by 2: $24 \times 2 = 48$

Next number: 48

Data Trend

EXAMPLE 2 The value of Scott Computer stock has risen as shown. If this pattern continues, what will the stock's value be in July?

Month:	April	May	June	July
Value:	$55\frac{1}{4}$	$55\frac{1}{2}$	$55\frac{3}{4}$	?

STEP 1 Find the common link.

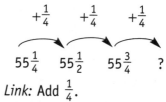

Link: Add $\frac{1}{4}$.

STEP 2 Find the predicted July value.

Add: $55\frac{3}{4} + \frac{1}{4} = 56$

Predicted July value: 56

Comparing Patterns

Sometimes two or more patterns can be found in a problem. Compare these patterns as rows on a table.

Laura plans to buy one of two used cars. The Honda costs $12,000 and will depreciate (lose value) at the rate of about $1,500 each year. The Ford costs $13,500 and will depreciate at the rate of about $2,000 per year. In how many years will the value of the cars be the same?

Write a table where the rows are the *depreciation patterns* described in the problem. Compare the rows year by year until the two values become equal.

Car	Value Now	After Year 1	After Year 2	After Year 3
Honda	$12,000	$10,500	$9,000	$7,500
Ford	$13,500	$11,500	$9,500	$7,500

ANSWER: The cars will have the same value ($7,500) after **3 years.**

Identify the patterns.

1. What is the next number in this series?

   $5, 8\frac{1}{2}, 12, 15\frac{1}{2}, \ldots$

2. Find the next term in the following series.

   $2, 6, 18, 54, \ldots$

3. Which series below is described by the following *number-to-number* rule "times two and subtract 3"?

   a. $5, 7, 14, 20, \ldots$
   b. $5, 7, 11, 19, \ldots$
   c. $5, 8, 13, 17, \ldots$
   d. $5, 8, 15, 19, \ldots$
   e. $5, 7, 14, 23, \ldots$

4. Choose the *alternate-number* rule that best tells how the following series is changing. $6, 12, 9, 15, 12, 18, 15, \ldots$

   a. add 6, add 3
   b. subtract 6, add 3
   c. add 4, subtract 3
   d. add 6, subtract 3
   e. add 3, subtract 6

Use the tables to compare patterns.

5. Pam has her choice of two part-time jobs. The job at The Valley Restaurant (TVR) pays $15,000 a year with yearly raises of $750. The job at Norman's Office Supplies (NOS) pays $13,650 a year with yearly raises of $1,200. In how many years will the annual salaries be equal? (Write the *salary patterns* on the table shown below. Compare year by year.)

Starting Salary	After 1 Year	After 2 Years	After 3 Years	After 4 Years	After 5 Years
TVR: $15,000					
NOS: $13,650					

6. As shown in the table below, Plant A and Plant B had different heights when growth measurements were started. If the weekly growth rate of each plant stays constant, at the end of which week will the two plants be the same height? (Write the *growth patterns* on the table below. Compare week by week.)

Starting Height	After 1 Week	After 2 Weeks	After 3 Weeks	After 4 Weeks	After 5 Weeks
Plant A: $6\frac{3}{4}$ in.	$7\frac{1}{2}$ in.				
Plant B: $6\frac{1}{4}$ in.	$7\frac{1}{8}$ in.				

Data Highlight: Reading a Pictograph

A **pictograph** uses pictures, or symbols, to graph information. The value of a symbol is defined in a key on the graph.

To find the total value of a line of symbols, multiply the number of symbols by the value of a single symbol. Half of a symbol has half the value of a complete symbol.

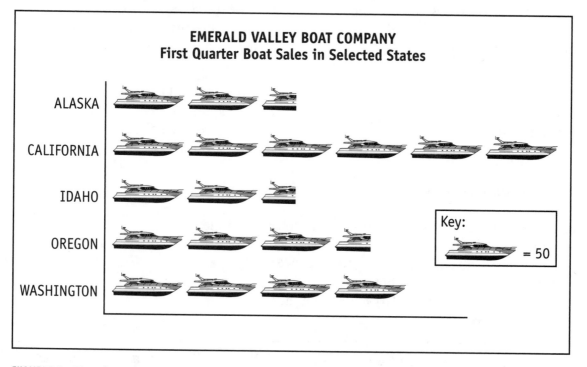

EMERALD VALLEY BOAT COMPANY
First Quarter Boat Sales in Selected States

ALASKA

CALIFORNIA

IDAHO

OREGON

WASHINGTON

Key: = 50

EXAMPLE 1 For the sales period shown, how many boats did Emerald Valley sell in Oregon?

> **STEP 1** Count the number of boat symbols to the right of Oregon. $3\frac{1}{2}$
>
> **STEP 2** Multiply $3\frac{1}{2}$ by 50, the value of each symbol.

ANSWER: 175 boats

EXAMPLE 2 What is the average number of boats sold per state for the five states listed?

> **STEP 1** Count the total number of boat symbols shown for the five states. $18\frac{1}{2}$
>
> **STEP 2** Divide: $18\frac{1}{2} \div 5 = 3\frac{7}{10}$ symbols per state
>
> **STEP 3** Multiply: $3\frac{7}{10} \times 50 = 185$ boats per state

ANSWER: An average of 185 boats was sold per state.

For problems 1–8, refer to the following pictograph.

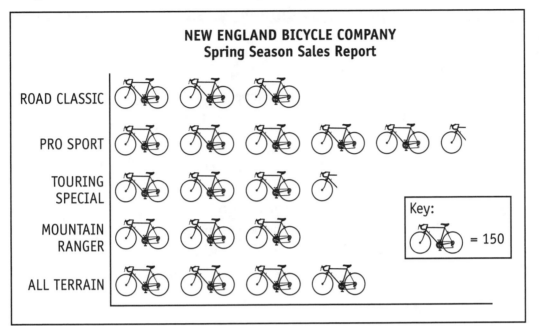

1. Which bicycle model sold the most during the spring season?

2. About how many more Pro Sports were sold this spring season than All Terrains?

3. About how many Mountain Ranger bicycles did the New England Bicycle Company sell during the spring season?

4. If Road Classics sell for $328, about how much sales revenue was received during the spring season from the sale of this model?

5. Which two bicycle models sold about the same number during the spring season?

6. Approximately how many bicycles did the New England Bicycle Company sell in all during its spring season?

7. If the spring season lasts 3 months, what is the *average* number of sales per month of the Pro Sport model?

8. What is the *ratio* of the number of Road Classics sold to the number of Touring Specials sold?

Solve the following problem.

9. The All Terrain model sells for $295, and the Pro Sport model sells for $445. What more do you need to know to determine how much more profit is made on each Pro Sport sale than on each All Terrain sale?

   a. the prices of each of the other three models of bicycles
   b. the total yearly sales of Pro Sport and All Terrain models
   c. the total yearly sales of all five bicycle models
   d. the cost of making each Pro Sport and each All Terrain model
   e. the weight of both the Pro Sport and All Terrain models

Fractions Review

Solve the problems below.

1. A pole $8\frac{7}{8}$ inches long is laid end to end with a second pole that is $14\frac{1}{4}$ inches long. Which of the following is the best estimate of the combined length of the two poles?

   a. 21 in.   **b.** 22 in.   **c.** 23 in.   **d.** 24 in.   **e.** 25 in.

2. A $12\frac{7}{8}$-ton railroad car is loaded with 18 pickup trucks each weighing $2\frac{1}{4}$ tons. Which expression below gives the best estimate of the combined weight in tons of the railroad car and the pickups?

   a. $18 \times (13 - 2)$ **c.** $(13 + 18) \times 2$ **e.** $18 \times 2 - 13$
   b. $12 + 18 \times 2$ **d.** $13 + 18 \times 2$

3. Stacey grew $1\frac{1}{4}$ inches two years ago, $1\frac{3}{8}$ inches last year, and $\frac{7}{8}$ inch so far this year. How many inches has Stacey grown during all this time?

4. The value of James Company stock fell from $16\frac{1}{2}$ to $14\frac{7}{8}$ between June 1 and June 15. How many points did the stock value drop during the 2-week period?

5. After driving half of the distance from Busby to Elmwood, how many miles is Kathleen from Elmwood?

   a. 34   **c.** 84   **e.** 126
   b. 66   **d.** 110

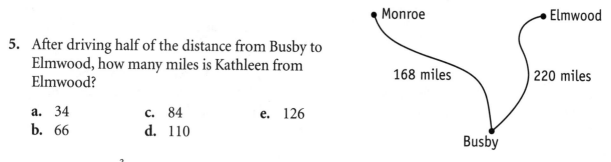

6. Denice mixes $\frac{3}{8}$ pint of thinner into each gallon of stain she uses. How many whole pints of thinner will Denice need for a house requiring 13 gallons of stain?

7. Bryan, a jewelry maker, uses $\frac{5}{16}$ ounce of gold for each ring he designs. How many complete rings can Bryan make when his gold supply is down to $2\frac{5}{8}$ ounces?

8. Of every 10 people who order burgers at Chuck's Place, 5 order single burgers, 3 order double burgers, and 2 order cheeseburgers. What is the ratio of single burger sales to cheeseburger sales?

   a. 1 to 5   **b.** 2 to 5   **c.** 2 to 1   **d.** 5 to 3   **e.** 5 to 2

9. During the first day of fall registration, 2,600 students enrolled at Walton College. Registration was completed in $6\frac{1}{2}$ hours. On average, how many students registered each hour?

   a. 400   **c.** 500   **e.** 600
   b. 450   **d.** 550

10. Willie is keeping a record of his daughter's growth. If her height keeps changing by the same amount each week, what will be her height at the end of the *4th week*?

Birth Height	Week 1	Week 2
$17\frac{3}{4}$ in.	$18\frac{5}{8}$ in.	$19\frac{1}{2}$ in.

11. Jonas found part of a bus schedule shown at the right. If it is now Tuesday, 11:50 A.M., when can Jonas catch the next bus at this station? (Assume that the next several departure times follow the pattern shown.)

    a. 11:45 A.M.   **c.** 12:10 P.M.   **e.** 12:35 P.M.
    b. 11:55 A.M.   **d.** 12:20 P.M.

Bus Schedule
Tuesday—Roosevelt Station
Departure Times

10:00 A.M.
10:25 A.M.
10:35 A.M.
11:00 A.M.
11:10 A.M.
11:35 A.M.

12. According to this pictograph, what is the ratio of the sales of Goldie dolls to the sales of Snazzle dolls?

    a. 3 to 2   **d.** 2 to 1
    b. 4 to 3   **e.** 3 to 1
    c. 5 to 4

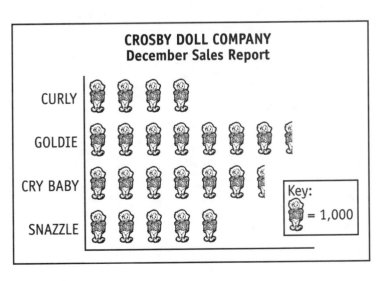

CROSBY DOLL COMPANY
December Sales Report

CURLY
GOLDIE
CRY BABY
SNAZZLE

Key: = 1,000

DECIMALS

Estimating: Building Confidence with Decimals

Suppose a test question requires you to solve the following:
26.04 − (4.83 + 6.075 + 1.6). Your answer choices are shown at the right.
This looks like a tough question! But estimation makes it easy. Simply
subtract the sum of 5 + 6 + 2 from 26, or 26 − 13 = 13. From the choices
given, you quickly (and correctly) choose **c. 13.535** as the answer.

a. 6.945
b. 9.005
c. 13.535
d. 20.945
e. 26.875

Estimation is useful in many situations. It can be a check to see that
an exact value makes sense, or it can help you choose an answer to a
multiple-choice question, as in the example above. Also, estimation can
help when you are not sure where to place the decimal point in an exact
answer.

Rounding Mixed Decimals

A **mixed decimal** is a whole number together with a decimal
fraction. When you read a mixed decimal, read the decimal point as
the word *and*. The mixed decimal 5.83 is read as "five *and* eighty-three
hundredths."

One way to estimate with mixed decimals is simply to round each
mixed decimal to the nearest whole number. Then do the indicated
math.

To round a mixed decimal to a whole number, look at the digit in
the tenths place.

> **Math Tip**
> When rounding to a
> whole number,
> concentrate only on
> the digit in the
> tenths place.

- If the digit is greater than or equal to 5, round up to the next
  larger whole number. Drop all the decimal digits.
- If the digit is less than 5, drop all the decimal digits. The whole
  number remains the same.

tenths place	tenths place	tenths place
14.6 rounds to 15	8.473 rounds to 8	6.54 rounds to 7
greater than or equal to 5	less than 5	greater than or equal to 5

14.6 ≈ 15   **8.473 ≈ 8**   **6.54 ≈ 7**

EXAMPLE 1 Estimate: 7.53 + 5.9

STEP 1 Round each mixed decimal to the
nearest whole number.
  7.53 ≈ 8   5.9 ≈ 6

STEP 2 Add the whole numbers.
  8 + 6 = 14

ANSWER: 7.53 + 5.9 ≈ 14

EXAMPLE 2 Estimate: 4.07 × 2.53

STEP 1 Round each mixed decimal to the
nearest whole number.
  4.07 ≈ 4   2.53 ≈ 3

STEP 2 Multiply the whole numbers.
  4 × 3 = 12

ANSWER: 4.07 × 2.53 ≈ 12

Round each decimal to the nearest whole number or dollar.

1. **a.** 3.87 meters ≈   **b.** $9.83 ≈   **c.** 7.13 centimeters ≈

2. **a.** $24.17 ≈   **b.** 5.75 ounces ≈   **c.** 29.2 miles per gallon ≈

Estimate an answer for each problem below.

	Estimate		Estimate		Estimate
3. $7.81 + 2.96		8.4 − 5.725		$9.27 × 19.5	

	Estimate		Estimate		Estimate
4. 9.75 × 0.95		2.2$\overline{)16.434}$		3.14$\overline{)42.076}$	

Use an estimate to help you decide where the decimal point belongs in each answer.

5. $8.75 + 6.392 + 1.4 = 1\ 6\ 5\ 4\ 2$

6. $13.37 \times 4.5 = 6\ 0\ 1\ 6\ 5$

7. $23.97 - (2.875 + 6.01) = 1\ 5\ 0\ 8\ 5$

8. $38.164 \div 4.7 = 8\ 1\ 2$

Estimate an answer to each problem. Then use your estimate to help you choose the correct answer from the choices given.

9. Three metal strips are placed side by side. How wide are they together if the individual widths are 6.95 cm, 9.1 cm, and 7.325 cm?

    a. 14.985   **b.** 17.245   **c.** 20.185   **d.** 23.375

10. Jade pays her neighbor's children $6.89 per hour to work in her garden. How much will Jade pay after the children have worked 41.75 hours?

    a. $231.74   **b.** $287.66   **c.** $329.85   **d.** $381.08

11. Ali bought 3.9 pounds of cole slaw for $0.96 per pound. How much change will Ali get when paying with a $10 bill?

    a. $5.08   **b.** $6.26   **c.** $7.92   **d.** $8.34

12. Blaine bought a 29.8-pound bag of potting soil for $4.88. About how many pounds of soil did she get per dollar?

    a. 6   **b.** 8   **c.** 10   **d.** 12

Rounding to Different Place Values

Place Values Beyond Thousandths

So far you have explored the first three decimal place values: tenths, hundredths, and thousandths. The place values continue beyond thousandths.

The First Six Decimal Place Values

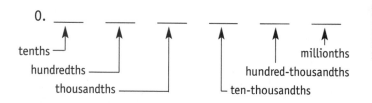

0. ___ ___ ___   ___ ___ ___

tenths
hundredths
thousandths
ten-thousandths
hundred-thousandths
millionths

Rounding to a Chosen Place Value

As discussed on the previous pages, rounding mixed decimals to the nearest whole number *before doing the math* is one way to estimate an answer. Rounding is also used to simplify an exact answer *after the math is done.*

To round a decimal fraction to a chosen place value, look at the digit to the *right* of the chosen place value.
- If the digit is greater than or equal to 5, round up. Drop all the digits to the right of the chosen place value.
- If the digit is less than 5, leave the digit in the chosen place value unchanged. Drop all digits to the right of the chosen place value.

Rounding to the Tenths Place (Nearest Tenth)

Check the digit in the hundredths place.

$0.75 \approx 0.8$
greater than or equal to 5

$1.43 \approx 1.4$
less than 5

$3.489 \approx 3.5$
greater than or equal to 5

Rounding to the Hundredths Place (Nearest Hundredth)

Check the digit in the thousandths place.

$0.298 \approx 0.30$
greater than or equal to 5

$1.271 \approx 1.27$
less than 5

$2.1453 \approx 2.15$
greater than or equal to 5

Rounding to the Thousandths Place (Nearest Thousandth)

Check the digit in the ten-thousandths place.

$0.9375 \approx 0.938$
greater than or equal to 5

$3.0573 \approx 3.057$
less than 5

$5.29975 \approx 5.300$
greater than or equal to 5

Round each decimal to the tenths place.

1. 0.46 ≈   0.52 ≈   0.85 ≈   0.235 ≈   0.912 ≈

2. 2.35 ≈   1.83 ≈   3.901 ≈   5.052 ≈   6.750 ≈

Round each decimal to the hundredths place.

3. 0.345 ≈   0.742 ≈   0.2753 ≈   0.892 ≈   0.3468 ≈

4. 3.625 ≈   4.052 ≈   8.500 ≈   6.9057 ≈   12.65805 ≈

Round each decimal to the thousandths place.

5. 0.3542 ≈   0.4627 ≈   2.93756 ≈   3.141593 ≈   27.00639 ≈

For problem 6, use the table to help you find the selling price.

6. Tami, a grocery clerk, used a calculator to multiply pounds (lb) by
   dollars per pound ($ per lb) to find the selling price of each
   product listed in the chart below. Her calculator answers are shown.

   Find the selling price of the packages by rounding each calculator
   answer to the nearest cent (hundredths place).

	Weight (lb)		Price ($ per lb)	Calculator Answer	Selling Price
Salad Bar	6.05	×	$1.79	10.8295	a._____
Fruit Salad	3.39	×	$2.09	7.0851	b._____
Cheese	2.80	×	$3.98	11.144	c._____
Chicken	1.85	×	$4.18	7.733	d._____
Hamburger	5.25	×	$4.80	25.2	e._____
Steak	4.15	×	$6.79	28.1785	f._____

(**Note:** A calculator does not display a dollar sign ($).)

Adding Decimals

To add decimals
- write the numbers in a column, lining up the decimal points (**Remember:** A whole number is understood to have a decimal point to the right of the ones digit.)
- use placeholding zeros if necessary to give all numbers the same number of decimal places (Extra zeros help keep columns in line.)
- add the columns
- place a decimal point in the answer directly below the decimal points in the problem

> **Math Tip**
> Write placeholding zeros at the *right* end of decimal fractions. In this way, you do not change their values.
>
> $3.5 = 3.50 = 3.500$

EXAMPLE 1 $3.243 + 2$

```
          ⌐——— Line up decimal points.
   3.243
 + 2.000  ← Write three placeholding zeros.
   5.243  ← Add the columns.
     └———— Place a decimal point in the answer.
```

ANSWER: 5.243

EXAMPLE 2 $1.9 + 0.75 + 0.125$

```
          ⌐——— Line up decimal points
   1.900  ← Write two placeholding zeros.
   0.750  ← Write one placeholding zero.
 + 0.125
   2.775  ← Add the columns.
     └———— Place a decimal point in the answer.
```

ANSWER: 2.775

 **Calculator Check**

Press Keys: [C] [3] [.] [2] [4] [3] [+] [2] [=]   (**Note:** You do not have to enter
Answer: [5.243]   placeholding zeros when using a calculator.)

Add. Use zeros as placeholders where necessary.

1.
```
   0.6        0.85        2.4       $0.56       $3.28
 + 0.3      + 0.62      + 1.7      +  0.49     +  1.39
```

2.
```
   0.83        2.4        7.63        9          3.875
   0.51        1.35       2           5.375      1.6
 + 0.125     + 0.96     + 1.875     + 3.65      + 5
```

Add. As your first step, line up the decimal points. Round each answer to the tenths place.

3. $8.75 + 6.3 + 4.325 \approx$ $2.875 + 1 + 0.25 \approx$ $7 + 4.5 + 2.125 \approx$

Subtracting Decimals

To subtract decimals
- write the numbers in a column, lining up the decimal points
- use placeholding zeros if needed to give all numbers the same number of decimal places
- subtract the columns just as you would subtract whole numbers
- place a decimal point in the answer directly below the decimal points in the problem

EXAMPLE $8 - 5.325$

STEP 1   Line up the decimal points.

$$\downarrow$$
$$8.$$
$$-\ 5.325$$

STEP 2   Use zeros as placeholders.

$$\downarrow\downarrow\downarrow$$
$$8.000$$
$$-\ 5.325$$
$$\overline{2.675}$$

ANSWER: 2.675

Calculator Check

Press Keys: [C] [8] [−] [5] [.] [3] [2] [5] [=]
Answer: [2.675]

Subtract. Use zeros as placeholders were necessary.

1.
0.9	0.74	5.7	13.9	$12.98
− 0.4	− 0.58	− 3.9	− 5.9	− 3.99

2.
0.875	6.8	4.56	21.9	30
− 0.35	− 0.95	− 2.275	− 8.583	− 16.735

Subtract. As your first step, line up the decimal points. Round each answer to the tenths place.

3. $2.25 - 1.9 \approx$ $9.85 - 4.7 \approx$ $18.2 - 9.75 \approx$ $35.5 - 25.875 \approx$

Subtract. Round each answer to the hundredths place.

4. $1.25 - 0.5625 \approx$ $7.4 - 5.875 \approx$ $12 - 9.625 \approx$ $23.5 - 8.9375 \approx$

Applying Your Skills

Solve the following word problems. They involve adding and subtracting decimals.

1. For a picnic, two tables are placed end to end. The first table is 2.75 meters long, and the second table is 2.5 meters long. What is the combined length of the two tables?

2. During the first half of the baseball season, Ramon's batting average was .284. During the second half of the season, his average improved by 0.13. Was Ramon's batting average over .300 by the end of the season?

3. The area of the United States is 3.62 million square miles, while the area of Mexico is 0.76 million square miles. By how many million square miles is the United States larger than Mexico?

4. Roger lives 8.75 miles from Bay City Airport and 12.4 miles from Kingsly Field Airport. How much closer is Roger to the airport at Bay City than to the one at Kingsly Field?

5. A Danish dining table has three leaves (removable center boards), the widths of which are shown in the diagram. Without the leaves, the table is 138 centimeters long. What is the length of the table with all three leaves?

    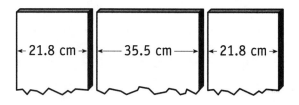

6. Cindy, a machinist, must cut a shaft so that the finished diameter (distance across) is 6.875 inches. If she starts with an 8-inch-diameter rod, how much will she need to reduce the diameter?

7. When he had the flu, Trong's temperature went from 98.6°F to 102.9°F. By how many degrees did Trong's temperature rise?

8. When he ran the 100-meter dash, Wade finished in exactly 12 seconds. How much slower is his time than the school record of 11.375 seconds?

9. Lynn has a bracket that is exactly 6 centimeters long. Use the diagram to tell, to the nearest millimeter, how much longer the bracket is than the bolt.

    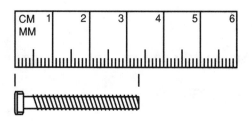

10. To pay for a $7 million convention center, the city council raised $1.24 million from the sale of bonds, $0.4 million from state economic development funds, and $2 million from private donations. How much more money does the council need to raise?

11. Carol is running in a 6.2-mile (10-km) road race. The first 1.25 miles are a slow uphill stretch, and the next 0.875 mile is a winding downhill section. The rest of the course is on level highway. How far will Carol be from the finish when she reaches the end of the downhill section?

12. To save postage, Myer is mailing several items in one box—but he must keep the total weight under 25 pounds. The box itself weighs 1.75 pounds, and the two items in it weigh 6.25 pounds and 7.5 pounds. Which equation below tells how many more pounds (*p*) Myer can place in the box?

    a. $p = 25 - (7.5 + 6.25 - 1.75)$
    b. $p = (7.5 + 6.25 + 1.75) - 25$
    c. $p = (7.5 + 6.25 - 1.75) - 25$

    d. $p = 25 - (7.5 + 6.25 + 1.75)$
    e. $p = 25 + (7.5 + 6.25 - 1.75)$

13. Last week Georgia earned $4.55 as her allowance. Her sister, Cara, earned $3.85. Her brothers earned $2.50 less than Georgia and Cara earned together. How much did the brothers earn last week?

14. In each pound of Halley's Mixed Nuts, by how much do the peanuts outweigh the cashews?

COMPOSITION OF HALLEY'S MIXED NUTS
(ingredients by weight in 1-pound mixture)

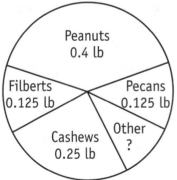

15. What part of a pound is represented by the word *Other* on the graph?

16. The table below compares the value (in millions of dollars) of WRT Company with that of CBA Company during 2 years of operation. If the present trend continues, after how many years will the two companies have the same value?

Starting Value	After 1 Year	After 2 Years	After 3 Years	After 4 Years	After 5 Years
WRT Company: 2.4	3.65	4.9			
CBA Company: 4.4	5.15	5.9			

Multiplying Decimals

To multiply decimals
- multiply the numbers
- count the number of decimal places in the numbers being multiplied to see how many decimal places are in the answer
- count off the number of decimal places, starting at the *right* of the answer; then place the decimal point

> **Math Fact**
> The number of decimal places in a number is the number of digits to the right of the decimal point. A whole number has zero decimal places.

EXAMPLE 1

```
  3.65  ←    2 decimal places
×    8  ←  + 0 decimal places
 2920        2 decimal places
```
↑ Place the decimal point so that the answer has two decimal places.

ANSWER: 29.20 = 29.2

EXAMPLE 2

```
  9.25  ←    2 decimal places
×  3.5  ←  + 1 decimal place
 4625        3 decimal places
2775
32375  Place the decimal point so that the
↑      answer has three decimal places.
```

ANSWER: 32.375

Calculator Check

Press Keys: | C | 9 | . | 2 | 5 | × | 3 | . | 5 | = |
Answer: | 32.375 |

Place a decimal point in each answer.

1.
```
  1.43  ←   2 places          8.5  ←  1 place          18.5  ←    1 place
×    5  ← + 0 places        ×0.4  ← + 1 place        × 0.06  ← + 2 places
  7 1 5     2 places         3 4 0    2 places         1 1 1 0    3 places
```

Multiply. Use estimation or a calculator to check your answers.

2.
| 4.5 | 0.23 | 15 | $0.45 | $3.29 |
| ×5 | × 6 | ×0.7 | × 8 | × 9 |

3.
| 2.16 | 10.5 | 1.52 | $4.08 | 0.91 |
| ×26 | ×3.4 | ×0.94 | ×5.5 | ×7.5 |

Multiply. As your first step, line up the digits. Round each answer to the tenths place.

4. $3.87 \times 5 \approx$ $0.56 \times 9 \approx$ $1.73 \times 0.86 \approx$ $0.75 \times 21 \approx$

Multiply. Round each answer to the hundredths place.

5. $\$6.48 \times 7.6 \approx$ $45.7 \times 3.65 \approx$ $31.25 \times 0.6 \approx$ $\$32.20 \times 4.63 \approx$

. . . .

Using Zero as a Placeholder

In some problems, it is necessary to write one or more zeros as placeholders before you can place the decimal point in the answer.

<u>EXAMPLE 3</u>

$$
\begin{array}{r}
0.5 \leftarrow \quad \text{1 decimal place} \\
\times \ 0.03 \leftarrow \ + \text{2 decimal places} \\
\hline
.\_15 \quad \quad \text{3 decimal places}
\end{array}
$$

Add 1 zero because there must be *three decimal places* in the answer.

ANSWER: 0.015

<u>EXAMPLE 4</u>

$$
\begin{array}{r}
0.87 \leftarrow \quad \text{2 decimal places} \\
\times \ 0.002 \leftarrow \ + \text{3 decimal places} \\
\hline
.\_ \ \_174 \quad \quad \text{5 decimal places}
\end{array}
$$

Add 2 zeros because there must be *five decimal places* in the answer.

ANSWER: 0.00174

. . . .

Multiply.

6.
$$
\begin{array}{r} 0.8 \\ \times\,0.02 \end{array}
\qquad
\begin{array}{r} 0.65 \\ \times\,3.2 \end{array}
\qquad
\begin{array}{r} 0.72 \\ \times\,0.46 \end{array}
\qquad
\begin{array}{r} 1.34 \\ \times\,0.018 \end{array}
\qquad
\begin{array}{r} 2.61 \\ \times\,0.053 \end{array}
$$

7. $0.4 \times 0.3 =$ $0.06 \times 0.9 =$ $0.0027 \times 8 =$ $0.25 \times 0.016 =$

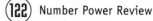

Dividing Decimals

Dividing by a Whole Number

To divide a decimal by a whole number
- place a decimal point in the quotient directly above its position in the dividend
- use one or more zeros as placeholders in the dividend as needed

EXAMPLE 1 Divide 31.2 by 6.

STEP 1 Place a decimal point in the quotient.

$$6\overline{)31.2}$$

Be sure the decimal points are lined up.

STEP 2 Divide as you do with whole numbers.

$$\begin{array}{r} 5.2 \\ 6\overline{)31.2} \\ -30 \\ \hline 1\,2 \\ -1\,2 \\ \hline 0 \end{array}$$

ANSWER: 5.2

EXAMPLE 2 Divide 8 into 0.184

STEP 1 Place a decimal point in the quotient.

$$8\overline{)0.184}$$

STEP 2 Since you can't divide 8 into 1, write a 0 above the 1. Now divide 8 into 18.

$$\begin{array}{r} 0.023 \\ 8\overline{)0.184} \\ -16 \\ \hline 24 \\ -24 \\ \hline 0 \end{array}$$

ANSWER: 0.023

Calculator Check

Press Keys: [C] [·] [1] [8] [4] [÷] [8] [=]
Answer: [0.023]

Divide. Use estimation or a calculator to check your answers.

1. $4\overline{)8.32}$ $5\overline{)0.755}$ $3\overline{)24.6}$ $12\overline{)0.684}$ $14\overline{)2.884}$

2. $6\overline{)0.444}$ $8\overline{)0.424}$ $6\overline{)0.774}$ $13\overline{)1.105}$ $27\overline{)0.0567}$

3. $4\overline{)\$9.16}$ $3\overline{)\$8.25}$ $5\overline{)\$12.75}$ $4\overline{)\$0.16}$ $8\overline{)\$0.24}$

Dividing by a Decimal

To divide a decimal by a decimal
- change the divisor to a whole number by moving its decimal point
- move the decimal point in the dividend an equal number of places to the right, adding placeholding zeros if needed
- place the decimal point in the quotient and divide

EXAMPLE 3 Divide $0.03\overline{)7.92}$

STEP 1 Move the decimal point in the divisor and dividend two places to the right.

$$0.03\overline{)7.92}$$

STEP 2 Divide

$$\begin{array}{r} 264. \\ 3\overline{)792.} \\ -6 \\ \hline 19 \\ -18 \\ \hline 12 \\ -12 \\ \hline 0 \end{array}$$

ANSWER: 264

EXAMPLE 4 Divide $0.004\overline{)16}$

STEP 1 Move the decimal point in the divisor three places to the right. Add three zeros to the dividend in order to move its decimal point three places to the right.

$$0.004\overline{)16.000}$$

STEP 2 Divide 4 into 16000.

$$\begin{array}{r} 4000 \\ 4\overline{)16000} \\ -16 \\ \hline 0000 \end{array}$$

ANSWER: 4,000

....

Divide.

4. $0.02\overline{)0.46}$ $0.14\overline{)0.42}$ $0.05\overline{)7.485}$ $1.2\overline{)20.64}$ $5.6\overline{)29.12}$

5. $0.03\overline{)1.5}$ $0.014\overline{)30.8}$ $0.012\overline{)0.144}$ $0.024\overline{)26.4}$ $0.016\overline{)0.208}$

6. $0.07\overline{)21}$ $0.03\overline{)18}$ $0.05\overline{)10}$ $2.5\overline{)50}$ $1.8\overline{)279}$

Using Placeholding Zeros When Dividing Whole Numbers

Use placeholding zeros to
- write a remainder as a decimal when dividing whole numbers
- change a proper fraction to an equivalent decimal

EXAMPLE 1 Divide 5 into 12 and write the remainder as a decimal.

Write the decimal point for the dividend 12. Then place a decimal point in the quotient.

$$\begin{array}{r} \downarrow \\ 2.4 \\ 5)\overline{12.0} \\ -10 \\ \hline 20 \\ -20 \\ \hline 0 \end{array}$$

Divide. ← Write a decimal point and add one placeholding zero.

ANSWER: 2.4

EXAMPLE 2 Change $\frac{3}{4}$ to a decimal.

Write the decimal point for the dividend 3. Then place a decimal point in the quotient.

Divide 4 into 3.

$$\begin{array}{r} \downarrow \\ .75 \\ 4)\overline{3.00} \\ -2.8 \\ \hline 20 \\ -20 \\ \hline 0 \end{array}$$

← Write a decimal point and add two placeholding zeros.

ANSWER: 0.75

Divide. Add enough zeros and write each remainder as a decimal.

1. $2)\overline{5}$ $\qquad$ $5)\overline{24}$ $\qquad$ $4)\overline{33}$ $\qquad$ $4)\overline{19}$ $\qquad$ $8)\overline{27}$

Divide to change each proper fraction to a decimal fraction.

2. $\frac{2}{4} =$ $\qquad$ $\frac{6}{8} =$ $\qquad$ $\frac{14}{20} =$ $\qquad$ $\frac{5}{8} =$ $\qquad$ $\frac{7}{16} =$

Sometimes when you divide two numbers, the quotient shows a repeating pattern. For example: $\frac{5}{11} = 0.454545...$ The result is called a **repeating decimal**.

> You can use three dots or a bar to show a repeating decimal.
> $\frac{1}{3} = 0.33...$
> $\frac{1}{3} = 0.\overline{3}$

Each fraction below has a repeating decimal equivalent. Divide. Then round each answer to the hundredths place.

3. $\frac{1}{3} \approx$ $\qquad$ $\frac{2}{3} \approx$ $\qquad$ $\frac{1}{6} \approx$ $\qquad$ $\frac{5}{6} \approx$ $\qquad$ $\frac{5}{9} \approx$

Multiplying or Dividing by 10, 100, or 1,000

To *multiply* a number by 10, 100, or 1,000, use one of these shortcuts.
You may need to add one or more placeholding zeros *at the right*.

To multiply by 10, move the decimal point *one place* to the right.

$0.6 \times 10 = 6$

$7.5 \times 10 = 75$

To multiply by 100, move the decimal point *two places* to the right.

$0.85 \times 100 = 85$

— added 0

$14.7 \times 100 = 1,470$

To multiply by 1,000, move the decimal point *three places* to the right.

$0.325 \times 1,000 = 325$

added 0s

$24.6 \times 1,000 = 24,600$

To *divide* a number by 10, 100, or 1,000, use one of these shortcuts.
You may need to add one or more placeholding zeros *at the left*.

To divide by 10, move the decimal point *one place* to the left.

$0.4 \div 10 = 0.04$

$2.9 \div 10 = 0.29$

To divide by 100, move the decimal point *two places* to the left.

$20.6 \div 100 = 0.206$

— added 0

$5.2 \div 100 = 0.052$

To divide by 1,000, move the decimal point *three places* to the left

$126 \div 1,000 = 0.126$

— added 0s

$3.4 \div 1,000 = 0.0034$

Multiply or divide as indicated.

1. $0.5 \times 10 =$ $0.86 \times 10 =$ $10 \times 3.7 =$

2. $0.75 \times 100 =$ $100 \times 2.6 =$ $15.2 \times 100 =$

3. $0.8 \times 1,000 =$ $9.25 \times 1,000 =$ $1,000 \times 25.6 =$

4. $0.7 \div 10 =$ $6.5 \div 10 =$ $12.7 \div 10 =$

5. $28.7 \div 100 =$ $3.4 \div 100 =$ $0.5 \div 100 =$

6. $125 \div 1,000 =$ $67 \div 1,000 =$ $9.8 \div 1,000 =$

Applying Your Skills

Solve the following word problems. They involve multiplying and dividing decimals.

1. At Vern's Market, Ellie bought 6.1 pounds of fruit salad selling at $1.29 per pound. Which of the following is the best estimate of what Ellie will pay for this purchase?

   a. $6.00   **b.** $6.90   **c.** $7.80   **d.** $8.50   **e.** $9.00

2. While traveling through Oregon, Annick drove 301.6 miles on 14.5 gallons of gas. Knowing this, figure out Annick's average mileage (miles per gallon) on this trip. Round your answer to the nearest mile per gallon.

3. Yann packed 30 cans of smoked salmon in a large shipping container. Each can of fish weighs 2.875 pounds, and the container itself weighs 23.25 pounds. What is the total weight of the packed container? Round your answer to the nearest pound.

4. In the metric system, road distance is measured in kilometers. 1 kilometer is equal to 0.62 mile. Giving your answer in miles, how much shorter is 50 kilometers than 50 miles?

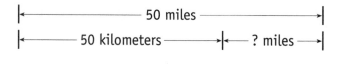

5. Kevin worked 19.6 hours of overtime last month, for which he was paid $243.98. Which expression below gives the best estimate of Kevin's overtime hourly pay rate?

   a. $200 × 10   **b.** $200 × 20   **c.** $200 ÷ 20   **d.** $250 ÷ 19   **e.** $250 ÷ 20

6. Jill earns time and three-quarters for each hour of overtime. Her overtime pay rate is found by multiplying her regular pay rate by 1.75. What is Jill's overtime rate if her regular rate is $6.80?

7. On Friday Josue sold four pickup loads of topsoil at his garden shop. What is the average weight of these four loads? Round your answer to the nearest hundredth ton.

   Load 1: 0.874 ton   Load 3: 0.73 ton
   Load 2: 1.05 tons   Load 4: 0.5 ton

8. A machinist placed four equal-size washers on a 1.125-inch bolt. The length of the uncovered part of the bolt is 0.875 inch. What is the exact width of each of the washers?

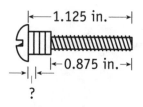

9. A box arrived at Kim's Import Food Store weighing 24.5 kilograms. The box contained 24 jars of pickled vegetables. When empty, the box weighs 1.6 kilograms. Which equation below tells how many kilograms (k) each jar weighs?

a. $k = (24.5 - 1.6) \div 24$

b. $k = (24.5 - 24) \div 1.6$

c. $k = (24.5 - 1.6) \times 24$

d. $k = (24.5 - 24) \times 1.6$

e. $k = (24.5 + 1.6) \div 24$

10. A digital scale shows weight in pounds as a mixed decimal number rounded to the nearest hundredth of a pound. Many markets and other shops use a digital scale to weigh a customer's purchase. What total price should the scale show?

4.73 lb	$3.89
Weight	$ per lb
Total Price	

11. At his office supplies store, Jackson got a package from UPS that weighed 53 pounds. In the package are 1,000 pens. If the box itself and the packing material have a total weight of 2.5 pounds, what is the approximate weight of each pen?

12. A 3-foot-wide bookshelf contains 20 books. How much room is left on the bookshelf if 10 of the books are 0.75 inch wide and 10 of the books are 1.125 inches wide?

13. Nadine needs to drill a hole that will allow a 0.325-inch-diameter wire to pass through. She wants the wire to fit as tightly as possible. Which of the three drill bits shown should she use?

Bit #	Diameter
1	$\frac{5}{16}$ inch
2	$\frac{11}{32}$ inch
3	$\frac{3}{8}$ inch

14. In Fremont, the cost of electric power is $0.0956 per kilowatt-hour. If, during the month of January, the Smith family used 3,000 kilowatt-hours of electric power, how much was their electric bill for that month? Write your answer showing dollars and cents.

15. Shaun is going to cut an 8-meter-long pipe into seven pieces. First, he will cut off a piece 3.28 meters long; then he will cut the remaining piece into six pieces of equal length. Circle the letter for the expression that gives the length of each of these shorter pieces.

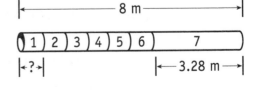

a. $\dfrac{8 - 3.28}{7}$

b. $\dfrac{7 - 3.28}{7}$

c. $\dfrac{8 - 3.28}{6}$

d. $\dfrac{7 - 3.28}{6}$

e. $\dfrac{6 - 3.28}{8}$

Data Highlight: Reading a Line Graph

A **line graph** gets its name from a line or lines that it uses to connect graphed data points. Numerical values are read along vertical and horizontal lines called **axes.**

The graph below represents the average oxygen consumption rate measured while jogging 12 minutes per mile.

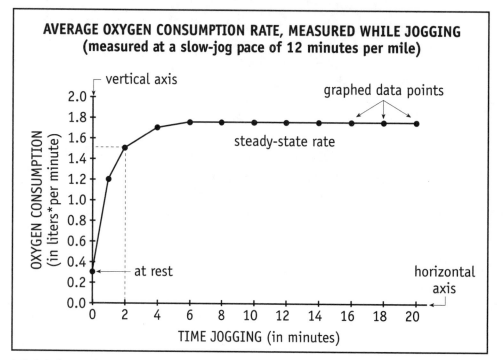

AVERAGE OXYGEN CONSUMPTION RATE, MEASURED WHILE JOGGING
(measured at a slow-jog pace of 12 minutes per mile)

*slightly larger than one quart

EXAMPLE 1   What is the approximate oxygen consumption rate of an average person at rest?

A person at rest would have spent 0 minutes jogging.

Find 0 on the *horizontal* axis (the axis that runs from left to right). Now read the data point indicated on the *vertical axis* (the axis that runs up and down).

ANSWER: A person at rest consumes about 0.3 liters of oxygen per minute.

EXAMPLE 2   What is the approximate oxygen consumption rate of an average person after 2 minutes of jogging?

STEP 1   Locate the graphed data point that is directly above the 2 on the horizontal axis.

STEP 2   Read the value on the vertical axis that is directly to the left of this data point.

ANSWER: After 2 minutes of jogging, a person consumes about 1.5 liters of oxygen per minute.

For problems 1 and 2, refer to the line graph on page 128.

1. Determine the approximate oxygen consumption rate of an average person after 4 minutes of jogging.

2. What is the approximate ratio of an average person's steady-state rate of oxygen consumption to his or her rate while at rest?

For problems 3–6, refer to the line graph below.

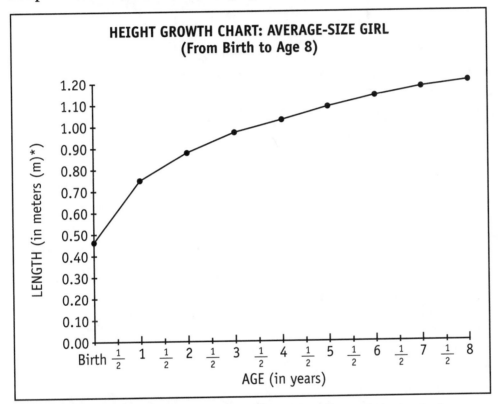

*When a length is written in meters, the first two decimal places are centimeters (cm). For example, 0.30 m = 30 cm, and 1.10 m = 110 cm.

3. Between which two birthdays does an average-size girl reach the height of 1 meter?

4. Approximately how much height (in cm) does a young girl gain between her first and third birthdays?

5. During which year does a baby girl's height increase most rapidly?

Decimals Review

Solve the problems below.

1. Anna bought 2.94 pounds of cheese for $3.89 a pound. What computation can you do to get a quick estimate of Anna's change if she pays with a $20 bill?

 a. $20 − ($4 × 2)   **c.** $20 − ($4 × 3)   **e.** $20 + ($3 × 2)
 b. $20 + ($4 × 3)   **d.** $20 − ($3 × 2)

2. Jeff delivers furniture for Mill's Furniture Store. On Saturday, Jeff drove 4.6 miles to the warehouse to pick up a bedroom set. He then drove 18.9 miles to the Malloys' to deliver it. He then drove 13 miles back to Mill's Furniture. How many miles did Jeff drive for this delivery?

3. Jacque's new skis are 1.87 meters long. How much shorter are these skis than the 2-meter skis he used last year?

4. What is the difference in length of these two nails?

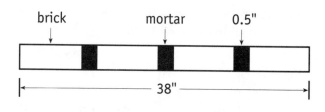

1.1875"   0.875"

5. In the metric system, small weights are measured in grams. (28.4 g ≈ 1 oz) Determine the weight in grams of an 8.75-ounce can of tuna.

6. A batting average is figured by dividing the number of hits a player gets by the number of times he or she bats. Last year Carmine was at bat 80 times and got 25 hits. To the nearest thousandth, what was Carmine's batting average?

7. Joni paid $16.15 for a 7.5-foot piece of pine molding. To the nearest cent, how much did Joni pay per foot?

8. Which expression can be used to find the length of each brick?

 a. $\dfrac{38'' - 0.5''}{4}$ **d.** $\dfrac{38'' - (4 \times 0.5'')}{4}$

 b. $\dfrac{38'' - 0.5''}{3}$ **e.** $\dfrac{38'' + (3 \times 0.5'')}{4}$

 c. $\dfrac{38'' - (3 \times 0.5'')}{4}$

9. What fraction below has a decimal equivalent that is a repeating decimal?

   a. $\frac{1}{8}$ b. $\frac{1}{5}$ c. $\frac{1}{4}$ d. $\frac{1}{3}$ e. $\frac{1}{2}$

10. One hundred tiles were lined up in a row. Each tile was 9 inches wide, and a space of 0.25 inches was left between tiles. To the nearest foot, how long is this row of tiles?

11. At the Paint Place, custom-made interior wall paint is on sale for $11.79 a quart. Custom-made interior enamel is on sale for $6.19 a pint. Which expression represents the best estimate of the cost of buying 4 quarts of custom-made interior wall paint and 3 pints of custom-made interior enamel?

   a. $3(\$6) + 4(\$12)$ c. $(3 + 4)(\$6 + \$12)$ e. $3(\$6) + 4(\$11)$
   b. $3(\$12) + 4(\$6)$ d. $3(\$5) + 4(\$11)$

12. Ellie is using a digital scale to measure the weight of each of three packages. The first package weighs 3.1 lb, the second weighs 4.25 lb, and the third weighs 5.166 lb. What is the average weight of these packages?

13. What is the speed in kilometers per hour of a car traveling 55 mph? (1 mi ≈ 1.6 km)

14. Zoe's calculator displays 8 digits, but it does not display unnecessary zeros and it does not round the right-most digit. If Zoe divides 14 by 3 on her calculator, what answer will be displayed?

15. What was the temperature in New York City at 9:00 P.M.?

16. About how many degrees Celcius did the temperature increase between 6:00 A.M. and 9:00 A.M.?

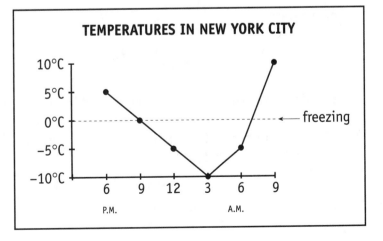

TEMPERATURES IN NEW YORK CITY

PERCENTS

Identifying Numbers in a Percent Problem

A percent problem involves three important numbers: the **percent,** the **whole,** and the **part.**

EXAMPLE 1

EXAMPLE 1

25%	of	$100	is	$25.
percent		whole		part

EXAMPLE 2

15	is	50%	of	30.
part		percent		whole

A percent problem asks you to find one of these three numbers when you know the other two.

> **Math Tip**
> In a percent problem, the *whole* always follows the word *of.*

- Find the *part* when you know the percent and the whole.
  What is 18% of $50?
  percent = 18%, whole = $50, part = ?

- Find the *percent* when you know the whole and the part.
  What percent of 75 is 15?
  whole = 75, part = 15, percent = ?

- Find the *whole* when you know the percent and the part.
  If 20% of a number is 14, what is the number?
  percent = 20%, part = 14, whole = ?

Identify the percent, the whole, and the part in each problem below.

1. 20% of 90 is 18.

   a. percent = _____

   b. whole = _____

   c. part = _____

2. $4.80 is 60% of $8.00.

   a. percent = _____

   b. whole = _____

   c. part = _____

Circle the letter for what you are asked to find in each problem below.

3. 30% of a number is 60. What is this number? You have to find the _____.

   a. percent **b.** whole **c.** part

4. 15 is what percent of 75? You have to find the _____.

   a. percent **b.** whole **c.** part

The Percent Circle

Each type of percent problem is solved by multiplication or by division. To remember whether to multiply or to divide, use a memory aid called the **percent circle.** The symbols P, %, and W label the three parts of the percent circle.

PERCENT CIRCLE

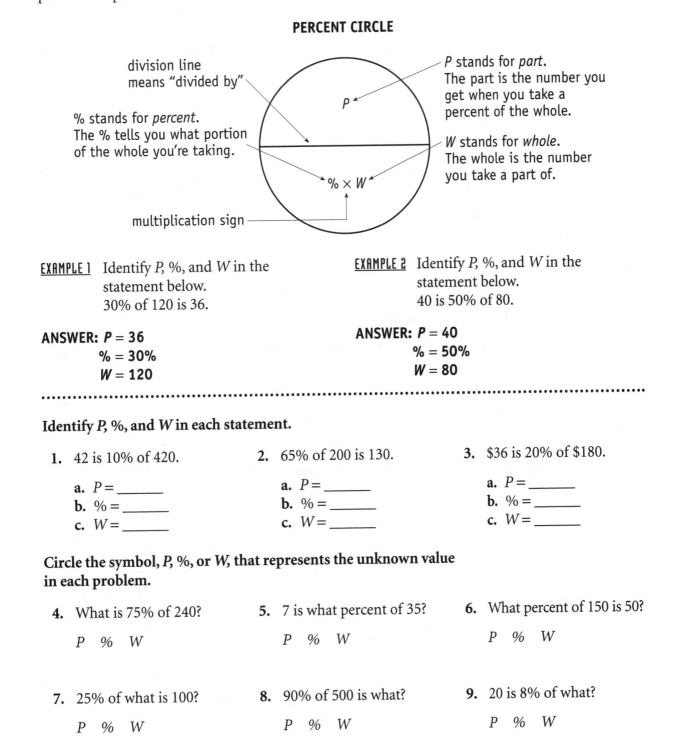

division line means "divided by"

% stands for *percent.* The % tells you what portion of the whole you're taking.

multiplication sign

P stands for *part.* The part is the number you get when you take a percent of the whole.

W stands for *whole.* The whole is the number you take a part of.

EXAMPLE 1   Identify P, %, and W in the statement below.
30% of 120 is 36.

ANSWER: P = 36
% = 30%
W = 120

EXAMPLE 2   Identify P, %, and W in the statement below.
40 is 50% of 80.

ANSWER: P = 40
% = 50%
W = 80

Identify P, %, and W in each statement.

1. 42 is 10% of 420.

  a. P = _____
  b. % = _____
  c. W = _____

2. 65% of 200 is 130.

  a. P = _____
  b. % = _____
  c. W = _____

3. $36 is 20% of $180.

  a. P = _____
  b. % = _____
  c. W = _____

Circle the symbol, P, %, or W, that represents the unknown value in each problem.

4. What is 75% of 240?

   P   %   W

5. 7 is what percent of 35?

   P   %   W

6. What percent of 150 is 50?

   P   %   W

7. 25% of what is 100?

   P   %   W

8. 90% of 500 is what?

   P   %   W

9. 20 is 8% of what?

   P   %   W

Using the Percent Circle

To use the percent circle

- cover the symbol of the number you're trying to find
- do the math indicated by the symbols that remain uncovered

EXAMPLE 1 Finding *part* of the whole.

If 15% of your $1,200 monthly paycheck is used for your car payment, how much is your car payment?

STEP 1 Cover P (the part)—the number you're trying to find.

STEP 2 Read the uncovered symbols: $\% \times W$

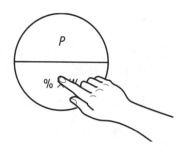

> **To find the part, multiply the percent by the whole.**

$P = \% \times W$ $0.15 \times \$1,200 = \textbf{\$180}$

EXAMPLE 2 Finding what *percent* a part is of a whole.

If $300 of your $1,200 monthly paycheck is used to pay rent, what percent is your rent payment?

STEP 1 Cover % (the percent)—the number you're trying to find.

STEP 2 Read the uncovered symbols: $\frac{P}{W}$ (This means $P \div W$.)

> **To find the percent, divide the part by the whole.**

$\% = \frac{P}{W}$ $\$300 \div \$1,200 = 0.25 = \textbf{25\%}$

EXAMPLE 3 Finding a *whole* when a part and a percent are given.

Suppose you buy a new 27″ color TV set and make a 10% down payment of $52. What is the price of the set?

STEP 1 Cover W (the whole)—the number you're trying to find.

STEP 2 Read the uncovered symbols: $\frac{P}{\%}$ (This means $P \div \%$.)

> **To find the whole, divide the part by the percent.**

$W = \frac{P}{\%}$ $\$52 \div 0.10 = \textbf{\$520}$

Fill in the percent circle. Then complete the three sentences below.

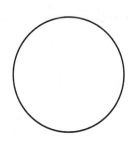

1. To find the part, _____.

2. To find the percent, _____.

3. To find the whole, _____.

First, circle the symbol that represents what you need to find. Next, decide whether you solve the problem by multiplication or division. *Do not solve these problems.*

4. Leona pays 25% of her monthly salary for rent. How much is Leona's rent payment if her monthly salary is $3,180?

   a. *P*   %   *W*

   b. _____ multiplication   _____ division

5. Seven of the 24 children in Mrs. Altman's first-grade class were home on Tuesday with colds. What percent of the class was sick that day?

   a. *P*   %   *W*

   b. _____ multiplication   _____ division

6. Suppose you make a down payment of $38 on a new sewing machine. If this down payment is 5% of the selling price, how much are you being charged for the machine?

   a. *P*   %   *W*

   b. _____ multiplication   _____ division

7. If you are charged a sales tax of $0.75 on a new $14.99 shirt, what percent is the sales tax of the cost of the shirt?

   a. *P*   %   *W*

   b. _____ multiplication   _____ division

8. According to a news report, 64% of the voters voted in favor of a school budget. If 28,680 people voted, how many voted in favor of the budget?

   a. *P*   %   *W*

   b. _____ multiplication   _____ division

Changing a Percent to a Fraction or a Decimal

The percent circle shows that percent problems are solved by multiplying or dividing by a percent. However, to multiply or divide by a percent, you first change the percent to a fraction or a decimal. Then you multiply or divide in the usual way.

On these next two pages, you want to strengthen your skill (studied earlier on pages 72–73) in changing a percent to either a fraction or a decimal.

Changing a Percent to a Fraction

To change a percent to a fraction
- write the percent as a fraction with a denominator of 100
- reduce the fraction if possible

EXAMPLE   Change 40% to a fraction.

STEP 1   Write 40% as 40 over 100. $40\% = \frac{40}{100}$

STEP 2   Reduce $\frac{40}{100}$ by dividing both $\frac{40 \div 20}{100 \div 20} = \frac{2}{5}$
the numerator and the denominator by 20.

ANSWER: $\frac{2}{5}$

> **Math Tip**
> The percents $33\frac{1}{3}\%$ and $66\frac{2}{3}\%$ occur so often in real-life problems that you may want to memorize the following facts.
> $33\frac{1}{3}\% = \frac{1}{3}$ $66\frac{2}{3}\% = \frac{2}{3}$

Changing a Percent to a Decimal

To change a percent to a decimal
- move the decimal point two places to the left, adding one or two zeros if necessary
(**Remember:** The decimal point of a whole number is understood to be at the right of the number, even though it may not be written.)
- drop the percent sign
- drop any unnecessary zeros

Percent	Move Decimal Point Two Places Left	Decimal
25%	25.	0.25 ↑ Write a leading zero.
60%	60. Add 1 zero.	0.60 = 0.6 ↑ Drop the unnecessary zero.
7%	07. Add 2 zeros.	0.07
0.5%	00.5	0.005

Write each percent as an equivalent fraction. Reduce the fraction if necessary.

1. 30%   50%   85% $66\frac{2}{3}\%$ $33\frac{1}{3}\%$

2. 1%   5%   9%   8%   6%

Write each percent as an equivalent decimal.

3. 25%   50%   75%   90%   7%

4. 5.5%   9.9%   8.5%   0.6%   0.1%

Solve each problem below.

5. At a Memorial Day sale, The Men's Shop reduced the prices of the items listed at the right. Write an equivalent fraction for each listed discount.

Sale Item	Discount	Fraction Off
a. Shirts	25%	_____
b. Sweaters	$33\frac{1}{3}\%$	_____
c. Jackets	10%	_____
d. Gloves	60%	_____
e. Boots	$66\frac{2}{3}\%$	_____
f. Scarves	75%	_____

6. When working with money, you can think of percent as meaning "cents per dollar." For example, 3% can be thought of as 3¢ per dollar. Why? Because $3\% = \frac{3}{100}$, and $3¢ = \frac{3}{100}$ of $1.

   Complete the table at the right, showing how many cents per dollar each percent represents.

Percent	Cents per Dollar
a. 5%	_____
b. 7%	_____
c. 12%	_____
d. 25%	_____
e. 50%	_____
f. 75%	_____
g. 90%	_____

Changing a Fraction or a Decimal to a Percent

Once in a while, you need to be able to change either a fraction or a decimal to a percent. You'll study these skills on the next two pages.

Changing a Fraction to a Percent

To change a fraction to a percent, multiply the fraction by 100%.

__EXAMPLE 1__ Change $\frac{3}{4}$ to a percent.

$$\frac{3}{\overset{}{\underset{1}{4}}} \times \frac{\overset{25}{\cancel{100\%}}}{1} = 75\%$$

ANSWER: 75%

__EXAMPLE 2__ Change $\frac{7}{8}$ to a percent.

$$\frac{7}{\overset{}{\underset{2}{8}}} \times \frac{\overset{25}{\cancel{100\%}}}{1} = \frac{175\%}{2} = 87.5\%$$

ANSWER: 87.5% or $87\frac{1}{2}$%

__EXAMPLE 3__ Change $\frac{2}{3}$ to a percent.

$$\frac{2}{3} \times \frac{100\%}{1} = \frac{200\%}{3} = 66\frac{2}{3}\%$$

ANSWER: $66\frac{2}{3}$%

Changing a Decimal to a Percent

To change a decimal to a percent

- move the decimal point two places to the right, adding a zero or two if necessary
- drop the decimal point if the percent is a whole number
- write a percent sign

> **Math Tip**
> You can think of changing a decimal to a percent as multiplying the decimal by 100%: multiply by 100 and add a percent sign. To multiply by 100, you move the decimal point two places to the right.

Decimal	Move the Decimal Point Two Places to the Right	Percent
0.2	0.20 ← Add a zero.	20% ← Drop the decimal point. Write a % sign.
0.75	0.75	75%
0.375	0.375	37.5%
1.25	1.25	125%
0.05	0.05	5% ← Drop the unnecessary zeros at the left.
3	3.00 ← Add two zeros.	300%

Write each fraction as a percent.

1. $\frac{1}{4}$ $\qquad$ $\frac{2}{5}$ $\qquad$ $\frac{3}{8}$ $\qquad$ $\frac{7}{10}$ $\qquad$ $\frac{1}{3}$ $\qquad$ $\frac{5}{6}$

Write each decimal as a percent.

2. 0.5 $\qquad$ 0.3 $\qquad$ 0.25 $\qquad$ 0.45 $\qquad$ 0.875

3. 0.275 $\qquad$ 0.06 $\qquad$ 1.75 $\qquad$ 4.5 $\qquad$ 6

Below is a partially completed chart of the most commonly used percents, decimals, and fractions. Complete this chart.

4.

Percent	Decimal	Fraction	Percent	Decimal	Fraction
	0.1				$\frac{3}{5}$
20%			$66\frac{2}{3}\%$		
	0.25			0.7	
		$\frac{3}{10}$			$\frac{3}{4}$
	$0.33\frac{1}{3}$		80%		
40%					$\frac{9}{10}$
		$\frac{1}{2}$		1	

Solve each problem.

5. At a Christmas sale, Jacob's Furniture advertised "All Chairs $\frac{1}{3}$ Off." Express this price reduction as a percent.

6. By April, $\frac{17}{20}$ of an apartment complex was completed.

   a. By April, what percent of the complex was completed?
   b. By April, what percent of the complex was not yet finished?

7. Citizen's Bank pays $0.045 per year for each dollar placed in a savings account earning simple interest. What percent savings rate is Citizen's Bank paying?

8. At Briggs High School, 0.625 of the teachers are women.

   a. What percent of the teachers at Briggs High School are women?
   b. What percent of the teachers at Briggs High School are men?

Finding the Part

Percent Circle: To find the part, multiply the percent by the whole.

How to Do It: Change the percent to a decimal or a fraction; then multiply.

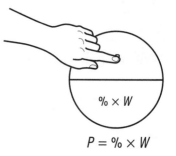

$\% \times W$

$P = \% \times W$

EXAMPLE An $80 radio is on sale for 25% off. How much can you save by buying the radio on sale?

You save 25% of $80.

Method 1

STEP 1 Change 25% to a decimal.
$25\% = 0.25$

STEP 2 Multiply $80 by 0.25.

$$\begin{array}{r} \$80 \\ \times\ 0.25 \\ \hline 400 \\ 1600 \\ \hline \$20.00 \end{array}$$

Methods 2

STEP 1 Change 25% to a fraction.
$25\% = \frac{25}{100} = \frac{1}{4}$

STEP 2 Multiply $80 by $\frac{1}{4}$.

$$\frac{\$80}{1} \times \frac{1}{4} = \frac{\$80}{4}$$

$$\frac{\$80}{4} = \$20$$

ANSWER: $20

ANSWER: $20

For most problems you may find method 1 (changing the percent to a decimal) to be easier. But when the percent is $33\frac{1}{3}\%$ ($\frac{1}{3}$) or $66\frac{2}{3}\%$ ($\frac{2}{3}$), method 2 (changing the percent to a fraction) is certainly easier!

Calculator Check

Press Keys: [C] [8] [0] [×] [2] [5] [%] [=]
Answer: [20.]

(**Note:** On some calculators, you must press [=] after [%].)

Find each part.

1. 25% of 68   15% of 200   50% of 128 $33\frac{1}{3}\%$ of 45

2. 6% of 75 $66\frac{2}{3}\%$ of $126   9.5% of 300   5.5% of $50

3. $8\frac{1}{2}\%$ of $70 $6\frac{1}{2}\%$ of $200   250% of 96   475% of $2,000
   (**Hint:** $8\frac{1}{2}\% = 8.5\%$.)   (**Hint:** 250% = 2.5.)

Solve each problem below.

4. In a state with a 6% sales tax, how much tax must be paid on a sweater that normally sells for $32?

5. Savings Mart is offering a 25% discount on all sport shirts in stock. What discount will be given on a sport shirt that normally sells for $30?

6. Randi's property tax bill is increasing by 3% this coming year. She now pays $890 a year in property tax.

   a. By how much is her property tax going to increase?
   b. How much property tax will she pay next year?

7. This morning, Lyle learned that he is getting a $5\frac{1}{2}$% raise at his part-time job. Before the raise, his salary was $320 per week.

   a. By how much is Lyle's weekly salary going to increase?
   b. What is Lyle's new weekly salary?

8. The newspaper reported that 8.5% of the 11,800 accidents in the state this year involved cars 15 years old or older. To the nearest hundred, how many accidents involved these older cars?

9. AGB Computer Company announced a $33\frac{1}{3}$% increase in sales this year. By how much did sales increase this year if last year's sales were $24 million?

10. How much do the Lyfords spend on food each month?

11. What is the Lyfords' monthly rent payment?

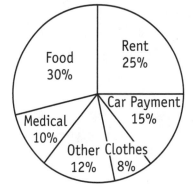

LYFORD FAMILY BUDGET
(monthly income: $1,500)

Food 30%
Rent 25%
Medical 10%
Other 12%
Clothes 8%
Car Payment 15%

Finding the Percent

Percent Circle: To find the percent, divide the part by the whole.

How to Do It: Write a fraction: $\frac{P}{W}$.

Reduce this fraction; then multiply by 100%.

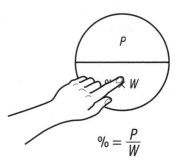

$$\% = \frac{P}{W}$$

EXAMPLE In Alicia's first-grade class, 6 of the 24 students are foreign-born. What percent of Alicia's class is foreign-born?

STEP 1 Write the fraction "part over whole" and reduce.

$$\frac{P}{W} = \frac{6}{24} = \frac{1}{4}$$

STEP 2 Multiply $\frac{1}{4}$ by 100%.

$$\frac{1}{\overset{1}{\cancel{4}}} \times \frac{\overset{25}{\cancel{100\%}}}{1} = 25\%$$

ANSWER: 25%

Calculator Check

Press Keys: | C | 6 | ÷ | 2 | 4 | % |

Answer: | 25. |

(**Note:** On some calculators, you must press | = | after | % |.)

Solve each problem below. Remember to reduce the $\frac{P}{W}$ fraction before multiplying by 100%.

1. What percent of 40 is 8?

2. $5 is what percent of $20?

3. What percent of 16 is 12?

4. 9 gallons is what percent of 27 gallons?

5. What percent of 8 pounds is 3 pounds?

6. 18 inches is what percent of 1 yard?
   (1 yard = 36 inches)

Solve each problem below.

7. On his math test, Walker got 48 questions correct out of 64. What percent of the test questions did Walker answer correctly?

8. When Hansen Real Estate sold Jim's condo, it charged him a commission of $4,500. If the sale price of the condo was $75,000, what percent real estate commission was he charged?

9. Yoshi has a part-time job. Out of her monthly gross pay of $1,200, Yoshi's employer withholds $180 for federal tax. Determine the percent of Yoshi's salary that is withheld for this tax.

10. The Sport House pays $12 for each basketball it buys. It then sells each ball for $20 and makes an $8 profit. Knowing this, figure out what percent markup (profit) is used by The Sport House. (**Hint:** The whole is the original price.)

11. Three years after Beth bought a used Honda for $14,500, its value decreased to $8,700.

    a. By how much did the value of the Honda drop in 3 years?
    b. What percent depreciation is this over the 3-year period?
    $$\left(\text{depreciation} = \frac{\text{value decrease}}{\text{original value}}\right)$$

12. During the fall, the weather report was correct only 26 times out of the 92 days that it predicted the next day's weather. Which expression below gives the percent of times it made *incorrect* predictions?

    a. $\frac{26}{92} \times 100\%$ c. $\frac{92 - 26}{92} \times 100\%$ e. $\frac{92}{92 - 26} \times 100\%$

    b. $\frac{92 + 26}{92} \times 100\%$ d. $\frac{92}{92 + 26} \times 100\%$

13. How much money is saved by buying each of the items listed above at the sale price?

    a. Sport coat:_____
    b. Wool pants:_____
    c. Wool sweater:_____
    d. Dress shoes:_____

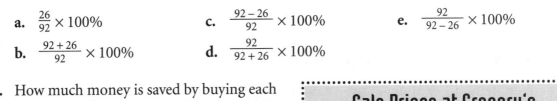

Sale Prices at Gregory's

Item	Original Price	Sale Price
Sport Coat	$80	$64
Wool Pants	$60	$45
Wool Sweater	$48	$32
Dress Shoes	$99	$33

14. What percent of the original price do you save by buying each listed item at the sale price?

    a. Sport coat:_____
    b. Wool pants:_____
    c. Wool sweater:_____
    d. Dress shoes:_____

Finding the Whole

Percent Circle: To find the whole, divide the part by the percent.

How to Do It: Change the percent to a decimal or a fraction; then divide.

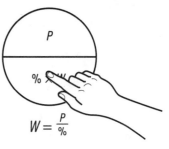

$$W = \frac{P}{\%}$$

EXAMPLE 30% of the people who applied this week for work at the new electronics plant were hired. If 60 people were hired, how many people submitted applications?

In other words, 60 is 30% of what number?

Method 1

STEP 1 Change 30% to a decimal.

$$30\% = 0.30 = 0.3$$

STEP 2 Divide 60 by 0.3

$$0.3\overline{)60.0} = 200$$

ANSWER: 200 people applied

Method 2

STEP 1 Change 30% to a fraction.

$$30\% = \frac{30}{100} = \frac{3}{10}$$

STEP 2. Divide 60 by $\frac{3}{10}$.

$$60 \div \frac{3}{10} = \frac{\overset{20}{\cancel{60}}}{1} \times \frac{10}{\underset{1}{\cancel{3}}} = \frac{200}{1}$$

$$\frac{200}{1} = 200$$

ANSWER: 200 people applied

Calculator Check

Press Keys: [C] [6] [0] [÷] [3] [0] [%]

Answer: 200.

(**Note:** On some calculators, you must press [=] after [%].)

Using either Method 1 or Method 2, solve these problems.

1. 15% of what number is 45%

2. 27 tons is 30% of what weight?

3. $36 is 20% of what amount?

4. $66\frac{2}{3}\%$ of what number is 150?
   (**Hint:** $66\frac{2}{3}\% = \frac{2}{3}$.)

Solve each problem below.

5. To pass his math test, Keith needs to get 60% of the questions correct. If Keith needs 42 correct answers to pass, how many questions are on the test?

6. Fifteen percent of Joyce's monthly income goes toward her car payment. If her car payment is $165, how much does Joyce make each month?

7. When he bought a new TV, Maurice made a 25% down payment of $87.50. What was the total price of the TV?

8. Dee was charged $4.50 during June on the unpaid balance on her Visa card. Her Visa company charges a 1.5% finance charge each month on her unpaid balance. How much was Dee's unpaid balance during June?

9. During the month of May, 40% of the babies born at St. Paul Hospital were boys. During that month, 90 girls were born at St. Paul.

   a. What *percent* of the babies born at St. Paul during May were girls?

   b. Using your answer from part a, figure out how many babies in all were born in St. Paul during May.

10. At a clearance sale, Kelli bought a blouse marked 70% off for $19.89. What percent of the original price did Kelli pay for the blouse?

11. Which expression below gives the best estimate of the original price of the blouse in problem 10?

   a. $\frac{\$20}{0.3}$ b. $\frac{\$20}{0.4}$ c. $\frac{\$20}{0.5}$ d. $\frac{\$20}{0.6}$ e. $\frac{\$20}{0.7}$

12. a. If 800 French students attend Madrid International College, what is the total student enrollment at the college?

   b. Using your answer from part a, determine how many North American students attend the college.

   c. What is the ratio of North American students to French students at the college?

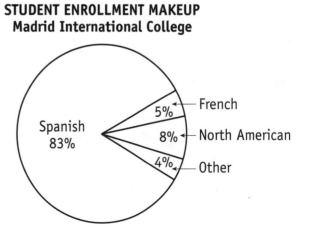

**STUDENT ENROLLMENT MAKEUP
Madrid International College**

Spanish 83%

5% — French

8% — North American

4% — Other

Problem Solver: Estimating with Easy Percents

In many percent problems (including many test questions), you do not need to compute an exact answer. An estimate will be close enough.

Be Familiar with These "Everyday Percent" Rules

The following rules come from changing everyday percents to fractions.

- 10% is the same as $\frac{1}{10}$ of a number. To find 10% of a number, divide by 10.

EXAMPLE   Find 10% of 80.   **Solution:** $80 \div 10 = \textbf{8}$

- 25% is the same as $\frac{1}{4}$ of a number. To find 25% of a number, divide by 4.

EXAMPLE   Find 25% of 48.   **Solution:** $48 \div 4 = \textbf{12}$

- $33\frac{1}{3}$% is the same as $\frac{1}{3}$ of a number. To find $33\frac{1}{3}$% of a number, divide by 3.

EXAMPLE   Find $33\frac{1}{3}$% of 30.   **Solution:** $30 \div 3 = \textbf{10}$

- 50% is the same as $\frac{1}{2}$ of a number. To find 50% of a number, divide by 2.

EXAMPLE   Find 50% of 82.   **Solution:** $82 \div 2 = \textbf{41}$

Work with Compatible Numbers

Compatible numbers are numbers that you can easily multiply or divide. Use compatible numbers whenever possible.

EXAMPLE 1   **Finding a part.**

What is 24% of 79?

Think: $24\% \approx 25\%$ *and* $79 \approx 80$

Rule: To find 25% of 80, divide 80 by 4.

   $80 \div 4 = 20$

Estimate: 20

EXAMPLE 2   **Finding a part.**

What is 11% of 412?

Think: $11\% \approx 10\%$ *and* $412 \approx 410$

Rule: To find 10% of 410, divide 410 by 10.

   $410 \div 10 = 41$

Estimate: 41

EXAMPLE 3   **Finding a percent.**

8 is what percent of 31?

Think: 8 is compatible with 32 ($32 \div 8 = 4$).

Write: $\frac{8}{31} \approx \frac{8}{32} = \frac{1}{4}$ and $\frac{1}{4} = 25\%$

Estimate: 25%

EXAMPLE 4   **Finding a whole.**

48% of a number is 92. What is the number?

Think: $48\% \approx 50\%$ *and* $92 \approx 90$

Write: $90 \div 50\% = 90 \div \frac{1}{2} = \frac{90}{1} \times \frac{2}{1} = 180$

Estimate: 180

Use estimation in each problem below. As a first step, decide whether you are asked to find the *part*, the *percent*, or the *whole*.

1. Joyce determined that 23% of her monthly income goes for rent. If her monthly income is $1,189, about how much is her rent?

2. When he bought the new gas range, Blake paid $50 down, or 9.5% of the sale price. What was the approximate sale price of the range?

3. Mari left a tip of $7.00 for a meal that cost $34.50. Approximately what percent tip did she leave?

4. According to a recent poll, $33\frac{1}{3}$% of the people polled said that they approved of the city's plan to expand the library. If 904 people were polled, about how many were in favor of expansion?

5. Starting next week, Josh will receive a raise of 5.2%. He now earns $7.00 per hour. Circle the letter of the expression below that most accurately describes the amount of Josh's raise.
   (**Hint:** 5.2% ≈ 5% = $\frac{1}{2}$ of 10%)

   a. a little less than 35¢ per hour
   b. a little more than 35¢ per hour
   c. a little less than 70¢ per hour

   d. a little more than 70¢ per hour
   e. a little less than $1.05 per hour

6. At Anson & Family Electronics, 67% of the employees are men. If Anson has 609 employees, how many of these employees are women?

7. Twenty-three percent of Stephen's monthly check goes to pay rent. If Stephen pays $345 per month for rent, how much is his monthly check?

8. The company Velma works for pays 75% of each employee's health care costs. Velma pays $36 a month for her health care.

   a. About how much does the company pay each month for health care coverage for each employee?
   b. About how much does the company pay each year for each employee for health care coverage?

9. At a restaurant, Sherry leaves a tip of about 15%. To do this, she estimates 10% of the meal cost. Then she takes half of this (5%) and adds that amount to her 10% estimate. Which expression below is the best estimate of the tip Sherry would leave for a $14.75 meal?

   a. ($15 ÷ 10) + ($15 ÷ 5)
   b. ($15 ÷ 10) − $\frac{1}{2}$($15 ÷ 10)
   c. ($15 × 10) + $\frac{1}{2}$($15 × 5)

   d. ($15 ÷ 10) + $\frac{1}{2}$($15 ÷ 10)
   e. ($15 ÷ 10) − ($15 ÷ 5)

Calculator Spotlight: Increasing or Decreasing an Amount

Many percent problems involve increasing or decreasing a whole by some percent. For problems of this type, a calculator greatly simplifies the steps involved. On these next two pages, you'll see just how useful a calculator can be in problem solving!

Suppose you want to find the purchase price of a car in a state where there is a sales tax.

Without a calculator, you can solve the problem by performing two steps.

- First, you multiply to find the amount of sales tax.
- Second, you add the sales tax to the selling price.

Using a calculator, you can combine these two steps into a single step.

EXAMPLE In a state with a 5% sales tax, what is the purchase price of a used car listed at $8,500?

STEP 1 Identify % and *W.*

% = 5%, *W* = $8,500

STEP 2 On your calculator, add $8,500 and 5% of $8,500 by pressing keys as follows.

	Press Keys	Display Reads
Clear the display.	C	0.
• Enter 8,500	8 5 0 0	8500.
• Press +	+	8500.
• Enter 5, then press %.*	5	5.
	%	8925.

ANSWER: $8,925

*On most four-function calculators, pressing % completes this calculation. On some calculators, though, you must press the = key after you press the % key.

Calculator Facts

When you press 8 5 0 0 + 5 %*, your calculator automatically adds 8,500 and 5% of 8,500.

If you want to subtract a percent from a whole (as you do in discount problems), press – instead of +.

Example: Pressing 2 0 0 – 6 %* subtracts 6% of 200 from 200 and gives the answer 188.

*On some calculators, you must press = to complete the calculation.

Solve the problems below using just paper and pencil. Then solve them using a calculator. Which way is easier for you? Do you see why a calculator is such a valuable tool?

Rate Increase: new amount = original amount + amount of increase

1. By changing jobs, Dave increased his monthly salary by 7%. What does Dave earn now if his previous job paid $1,230 per month?

2. This year, the Jacobsens' average monthly heating bill has been $86. However, experts predict a 4.5% rise in energy prices for next year. If they're right, how much (to the nearest dollar) can the Jacobsens expect to pay in monthly heating bills next year?

Rate decrease: new amount = original amount – amount of decrease

3. By turning their thermostat down to 68 degrees during the winter months, the Mondales hope to save 6% on their monthly heating bills. If they succeed, how much will the Mondales' heating bills average this winter if last winter they averaged $182.50?

4. To lose weight, Gregory has been advised to cut his daily calorie intake by 30%. Before starting this diet, Gregory consumed about 2,700 calories each day. While on the diet, what target level (to the nearest 100 calories per day) should Gregory try to attain?

Markup: selling price = store's cost + markup

5. Central Hardware places a 30% markup on each item it sells. If Central pays $13.50 for StrongArm Hammers, what price will Central charge its own customers for these hammers?

6. At Hermann's Men's Store, Hermann pays $59 for Paris Nights sweaters. Hermann then adds a 35% markup to his cost. What does Hermann charge his customers for these sweaters?

Discount: sale price = original price – amount of discount

7. Valley Appliances is marking down its appliances by 20%. What will be the sale price of a washer that normally sells for $289?

8. Salem Motors is offering a 15% discount on any used car in stock. Interested in a good deal, the Carmino family is looking at a used compact car that has a sticker price of $9,800. To the nearest hundred dollars, what will be the discounted price?

Life Skill: Understanding Simple Interest

Interest is money that you earn (or pay) for the use of money.

- If you deposit money in a savings account, the bank pays you interest for use of your money.
- If you borrow money, you pay interest to the lender for use of its money.

Simple interest is interest earned (or paid) on the amount of principal—the amount that is deposited or borrowed.

To compute simple interest, you use the simple interest formula.

In words: *i*nterest equals *p*rincipal times *r*ate times *t*ime

In symbols: $i = prt$ (which means $p \times r \times t$)

interest (*i*)	=	principal (*p*)*	×	rate (*r*)	×	time (*t*)
Expressed in dollars		Expressed in dollars		Expressed as a decimal		Expressed in years

*Be sure not to confuse the use of the letter *p* in the simple-interest formula with its use in the percent circle. In $i = prt$, *p* stands for *principal*. In the percent circle, *P* stands for *part*.

EXAMPLE 1 How much interest is earned on $600 deposited for 3 years in a savings account that pays 5% simple interest?

STEP 1 Identify *p, r,* and *t.*
$p = \$600, r = 5\% = 0.05, t = 3$ yr

STEP 2 Replace the letters in *prt* with the values given in Step 1.
$i = prt = \$600 \times 0.05 \times 3$

$$
\begin{array}{r} \$600 \\ \times\ 0.05 \\ \hline \$30.00 \end{array} \rightarrow \begin{array}{r} \$30 \\ \times\ 3 \\ \hline \$90 \end{array}
$$

ANSWER: $90

(**Note:** You can also write $5\% = \frac{5}{100}$.
In this case, you would multiply as follows:

$i = \$\overset{6}{600} \times \frac{5}{\underset{1}{100}} \times 3 = \$90$)

EXAMPLE 2 How much interest would you pay on $500 loaned to you at a simple-interest rate of 8.5% for 2 years?

STEP 1 Identify *p, r,* and *t.*
$p = \$500, r = 8.5\% = 0.085, t = 2$ yr

STEP 2 Replace the letters in *prt* with the values given in Step 1.
$i = prt = \$500 \times 0.085 \times 2$

$$
\begin{array}{r} \$500 \\ \times\ 0.085 \\ \hline 2500 \\ 40000 \\ \hline \$42500 \end{array} \rightarrow \begin{array}{r} \$42.50 \\ \times\ 2 \\ \hline \$85.00 \end{array}
$$

ANSWER: $85.00

Use the simple-interest formula to solve each problem.

1. How much would you earn on a deposit of $2,500 in 2 years if you were paid a simple-interest rate of 6%?

2. Suppose you deposit $800 in a savings account that earns 4.5% simple interest.

   a. How much interest will you earn in 2 years?
   b. What will be the total in your account at the end of the 2 years? (total balance = principal + interest)

3. The loan rates charged by States Bank are shown at the right. Suppose you borrow $8,000 to buy a used car. The bank agrees to let you pay the entire principal plus interest at the end of 3 years. What total amount will you owe the bank at the end of 3 years? (balance owed = principal + interest)

States Bank Simple-Interest Loans	
New Car	9.9%
Used Car	10.5%

Repayment Schedules

When you borrow money from a bank or credit union, you will usually repay it according to a **repayment schedule.** A repayment schedule lists interest rates, loan amounts, and monthly payments.

Use the repayment schedule to solve each problem.

Monthly Repayment Schedule

Rate	Loan Amount	24 months	36 months	48 months
10%	$5,000	$231	$161	$127
	$7,500	$346	$242	$191
12%	$5,000	$235	$166	$131
	$7,500	$353	$249	$197

Math Tip
In a repayment schedule, an interest charge is included as part of each monthly payment.

4. Suppose you take out a $7,500 loan at an interest rate of 10% for 24 months.

   a. What will be your monthly payment?
   b. How much money will you pay to the credit union during the 24 months?
   c. How much total interest will you pay for this loan? (total interest = total payments − $7,500)

5. Suppose you take out a $5,000 loan at an interest rate of 12% for 48 months.

   a. What will be your monthly payment?
   b. How much total interest will you pay for this loan?
   c. How much interest can you save on this $5,000 loan by paying it off in 36 months rather than in 48 months?

Computing Interest for Part of a Year

Interest is usually listed as a yearly rate. However, many deposits or loans involve a time period that is not a whole number of years.

When using the simple-interest formula for part of a year, write the time as either a fraction or a decimal—depending on which one makes the multiplication easier.

__EXAMPLE 1__ How much interest is earned on a $600 deposit if you are paid an interest rate of 6% and you leave your money in the bank for 8 months?

__STEP 1__   Identify p, r, and t.

$$8 \text{ mo} \downarrow$$

$$p = \$600, \, r = 6\% = \tfrac{6}{100}, \, t = \tfrac{8}{12} = \tfrac{2}{3}$$

$$\uparrow$$
$$12 \text{ mo} = 1 \text{ yr}$$

__STEP 2__   Multiply.

$$i = prt$$

$$= \$600 \times \tfrac{6}{100} \times \tfrac{2}{3}$$

$$= \tfrac{\overset{6}{\cancel{\$600}}}{1} \times \tfrac{\overset{2}{\cancel{6}}}{\underset{1}{\cancel{100}}} \times \tfrac{2}{\underset{1}{\cancel{3}}}$$

$$= 24$$

__ANSWER: $24__

__EXAMPLE 2__ How much interest will you pay on a loan of $700 borrowed for 1.5 years at a 14% interest rate?

__STEP 1__   Identify p, r, and t.

$$p = \$700, \, r = 14\% = 0.14, \, t = 1.5$$

__STEP 2__   Multiply.

$$i = prt$$

$$= \$700 \times 0.14 \times 1.5$$

$$= \$98 \times 1.5$$

$$= \$147$$

__ANSWER: $147__

Solve.

1. Write each time period as a proper fraction. Reduce if possible.

   4 months   6 months   9 months   11 months

2. Write each time period as a decimal. Round answers to the hundredths place.

   3 months   4 months   9 months   10 months

3. Write each time period as indicated.

   1 year 3 months   2 years 4 months   2 years 9 months
   improper fraction: _____   decimal: _____   improper fraction: _____

4. How much interest will you earn on a $400 deposit placed in a savings account for 9 months if you're paid at a simple-interest rate of 5%?

5. Suppose you take out a $650 loan. How much interest will you owe at the end of 8 months if you are charged a simple-interest rate of 10%?

6. At a bank that pays 4% simple interest, how much can you earn on a deposit of $900 in 18 months?

7. Suppose you put $1,000 in a savings account that pays 5% simple interest. What will be your balance in the account after 15 months? (balance = principal + interest)

8. You borrow $1,500 and are charged 14% simple interest. What total amount will you owe the lender after 1 year 9 months? (balance owed = principal + interest)

9. How much will it cost you to repay a $700 loan after 15 months if you are paying a simple-interest rate of 12%?

Credit-Card Interest Rates

Millions of people have a credit card. Many of these cards charge 1.5% per month interest on any unpaid balance. Often called "a small monthly finance charge," a 1.5% monthly rate is actually equal to an annual percentage rate (APR) of 18.9%!

Month	Unpaid Balance	Finance Charge
May	$80.00	
June	$120.00	
July	$200.00	
August	$350.00	

10. Suppose you have a Visa card that charges 1.5% per month on any unpaid balance.

   a. Calculate the monthly finance charge you would owe at the end of each month listed at the right. (finance charge = unpaid balance × 1.5%)

   b. Suppose you charged $400 on your Visa and made no monthly payments for 1 year. How much interest would you be charged for the year on this $400? (yearly interest = principal × APR × 1)

11. Suppose you are going to make a $600 purchase using a credit card. You plan to pay off the balance quickly and pay interest for only 1 month. Which of the two Visa cards below would be less expensive for you for this purchase?

   National Visa: Charges 1.5% per month
   Diamond Visa: Charges 1% per month plus a $2 monthly service charge

Data Highlight: Reading a Bar Graph

A **bar graph** gets its name from the bars that it uses to show data. These bars may be drawn vertically (up and down) or horizontally (across). You read a value for each bar by finding the number on the axis that corresponds to the end of the bar.

Usually, a bar graph is used to show a quick comparison of values. Exact values are often hard to obtain from a bar graph.

The bar graph below shows the average percent of fat found in several common foods.

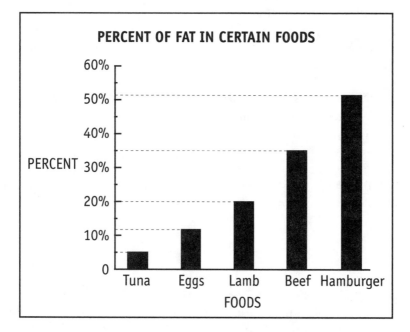

PERCENT OF FAT IN CERTAIN FOODS

EXAMPLE 1 What is the average percent of fat found in beef?

STEP 1 Locate the top of the bar labeled *Beef*.

STEP 2 Read the value on the vertical axis that is directly to the left of the top of this bar.

ANSWER: about 35%

EXAMPLE 2 What is the approximate ratio of the amount of fat found in hamburger to the amount found in tuna?

STEP 1 Determine the approximate percent of fat found in hamburger and tuna. hamburger ≈ 50% and tuna ≈ 5%

STEP 2 Determine the ratio. $\frac{\text{hamburger}}{\text{tuna}} \approx \frac{50\%}{5\%} = \frac{10}{1}$

ANSWER: about 10 to 1

For problems 1–3, refer to the bar graph on page 154.

1. Which of the listed foods contains the highest percent of fat?

2. Of the listed foods, which contain less than 30% fat?

3. **a.** What is the approximate percent of fat found in eggs?

   b. A large egg weighs about 2 ounces. Knowing this, determine approximately how much fat is contained in a serving of two large eggs. Express your answer to the nearest tenth of an ounce.

For problems 4–9, refer to the bar graph at the right.

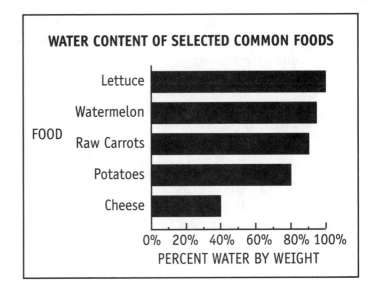

4. Approximately what percent of the weight of carrots is made up of water?

5. Which of the listed foods contain more than 90% water by weight?

6. To the nearest ounce, how much water is contained in a pound of carrots? (**Remember:** 1 pound = 16 ounces.)

7. How much more water is contained in 8 ounces of potatoes than in 8 ounces of cheese? Express your answer to the nearest ounce.

8. What is the approximate ratio of the percent of water in potatoes to the percent of water in cheese?

9. When you pay $0.21 per pound for watermelon, about how much are you paying per pound for water?

Percents Review

···

Solve the problems below.

1. A magazine reported that 35% of the people who attended the rally were women. If 133 women were there, which expression could you use to find how many people attended in all?

   a. $133 \div 0.35$ **b.** $0.35 \div 133$ **c.** $133 \div 35$ **d.** 0.35×133 **e.** 133×35

2. Which two have the same value as 40%?

   a. 4   **b.** 0.4   **c.** $\frac{2}{5}$ **d.** $\frac{4}{5}$

3. Fortrell Lumber Company announced that it would rehire $66\frac{2}{3}\%$ of the 291 workers it laid off last month. How many workers are going to be rehired?

4. Including a 5% sales tax, what is the total cost of the toaster?

$24.00

5. For selling 25 sets of Children's Fairy Tales, Lynn was paid a $200 commission. If her total sales receipts are $1,250, what percent commission does Lynn earn?

6. When she bought a stove, Ming made a 20% down payment of $82.50. What is the purchase price of Ming's new stove?

7. Clara placed $650 in a savings account that pays 4% simple interest. How much interest will Clara earn from this account if she leaves her money in for 18 months?

8. The Clothes House prices all skirts at $33\frac{1}{3}\%$ more than it pays for them. Which expression below shows how much customers are charged for a skirt that The Clothes House buys for $36?

   a. $\$36 - (\frac{1}{3} \times \$36)$ **c.** $\$36 - (\$36 \div \frac{1}{3})$ **e.** $\$36 + (\frac{2}{3} \times \$36)$

   b. $\$36 + (\frac{1}{3} \times \$36)$ **d.** $\$36 + (\$36 \div \frac{1}{3})$

9. A camera that usually sells for $290 is on sale for $240. Which expression below can be used to find the percent of *price decrease* of this camera during the sale?

a. $\frac{\$240}{\$290} \times 100\%$

b. $\frac{\$290 - \$240}{\$240} \times 100\%$

c. $\frac{\$290 + \$240}{\$240} \times 100\%$

d. $\frac{\$290 - \$240}{\$290} \times 100\%$

e. $\frac{\$290 + \$240}{\$290} \times 100\%$

10. What is the approximate trade-in value after 1 year of a car that cost $13,800 when new?

11. Approximately what was the original purchase price of a car that has a trade-in value of $7,500 when it is 3 years old?

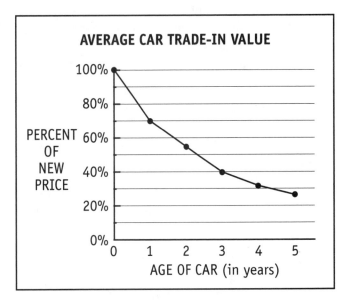

AVERAGE CAR TRADE-IN VALUE

PERCENT OF NEW PRICE (vertical axis: 0% to 100%)

AGE OF CAR (in years) (horizontal axis: 0 to 5)

12. Carla borrowed $2,960 from Lane Credit Union at a simple-interest rate of 10.25%. Which expression gives the best estimate of the total amount Carla will owe Lane at the end of 1 year 11 months?

a. $\$3,000 \times \frac{1}{10} \times 2$

b. $43,000 - (\$3,000 \times \frac{1}{10} \times 2)$

c. $\$3,000 + (\$3,000 \times \frac{1}{10} \times 2)$

d. $\$3,000 + (\$3,000 \times \frac{1}{100} \times 2)$

e. $\$3,000 + (\$3,000 \times 10 \times 2)$

13. If Benson County had 68,400 registered voters in 1988, how many of these people did *not* vote in the 1984 election?

14. In 1996, there were 66,400 Benson County residents who voted in the presidential election. How many total people were registered to vote in Benson County that year?

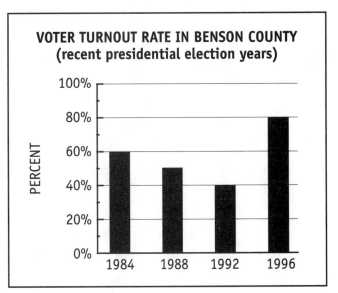

VOTER TURNOUT RATE IN BENSON COUNTY
(recent presidential election years)

PERCENT (vertical axis: 0% to 100%)

(horizontal axis: 1984, 1988, 1992, 1996)

DATA ANALYSIS AND PROBABILITY

Locating Data

This chapter discusses several special topics concerning **data,** which is information given as numbers, measurements, or other descriptions.

- **Data analysis** deals with interpreting data and drawing logical conclusions.
- **Probability** deals with making predictions based on laws of chance.

Interpolation

To **interpolate** is to *estimate* the value of a data point that lies *between* two given values.

<u>EXAMPLE 1</u> Referring to the table at the right, determine Jesse Baker's approximate weight at age 27 months.

Notice that 27 lies halfway between 24 and 30 months. Ask, "What weight is halfway between 22 and 26 pounds?"

The answer, **24 pounds,** is your best estimate of Jesse's weight at age 27 months.

Weight Growth Chart of Jesse Baker

Age (in months)	Weight (in pounds)
18	18
24	22
27 ←------------------→ ?	
30	26
36	30

Extrapolation

To **extrapolate** is to *estimate* the value of a data point that lies *outside the range* of your given data.

<u>EXAMPLE 2</u> Refer again to the table above. About what will Jesse weigh at age 42 months?

Notice that Jesse's weight pattern has increased 4 pounds every 6 months.

Ask yourself, "If this pattern continues, what will Jesse weigh in 6 more months—at age 42 months?" To estimate, add 4 to 30.

The answer, **34 pounds,** is your best estimate of Jesse's weight at 42 months.

Age (in months)	Weight (in pounds)	
18	18	
		+ 4 lb
24	22	
		+ 4 lb
30	26	
		+ 4 lb
36	30	
		+ 4 lb
42	?	

Use the following data to answer the problems.

1. **a.** Estimate the median weight of a 5'9" tall man of average build.

   b. Estimate the median weight of a 6' tall woman of average build.

Median Weight of Average-Build Adults (in pounds)		
Height	**Women**	**Men**
5'2"	115	124
5'4"	122	133
5'6"	129	142
5'8"	136	151
5'10"	144	159
6'		167

2. **a.** What was the approximate Death Valley temperature at 12:30 P.M. on July 8?

   b. If this temperature pattern continued, what did the temperature in Death Valley drop to by 8:00 P.M.?

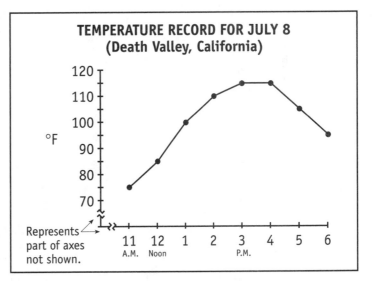

3. **a.** From the sales pattern shown, estimate the sales figures for the month of May.

   b. Assuming the sales pattern continues, about how many shoes can Delmont expect to sell in August?

Using More than One Data Source

Sometimes you may need to use more than one data source to determine needed information.

EXAMPLE Using the information given below, determine the amount of protein in a 6-ounce serving of Millie's Famous Sandwich Spread.

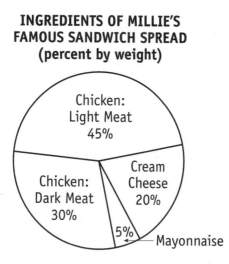

INGREDIENTS OF MILLIE'S FAMOUS SANDWICH SPREAD
(percent by weight)

Chicken: Light Meat 45%
Chicken: Dark Meat 30%
Cream Cheese 20%
5% Mayonnaise

Nutrition Information

	Protein*	Fat*
Chicken: Light Meat	9	1
Chicken: Dark Meat	8	2
Cream Cheese	2	10
Mayonnaise	0	11

*grams per ounce

STEP 1 Using information from the circle graph, determine how many ounces of each ingredient are in a 6-ounce serving.

ounces of ingredient = percent of ingredient × 6 ounces

Chicken: light meat = 45% × 6 = 2.7 ounces
Chicken: dark meat = 30% × 6 = 1.8 ounces
Cream cheese = 20% × 6 = 1.2 ounces
Mayonnaise = 5% × 6 = 0.3 ounce

STEP 2 Using information given in the table, compute the amount of protein in each ingredient.

amount of protein = protein per ounce × number of ounces

Chicken: light meat = 9 × 2.7 = 24.3 grams
Chicken: dark meat = 8 × 1.8 = 14.4 grams
Cream cheese = 2 × 1.2 = 2.4 grams
Mayonnaise = 0 × 0.3 = 0 grams

ANSWER: Total grams of protein = 24.3 + 14.4 + 2.4 + 0 = 41.1 ≈ 41 grams

A 6-ounce serving of Millie's Famous Sandwich Spread contains about **41 grams** of protein.

Use the circle graph and table on page 160 to solve problem 1.

1. Fill in the table at the right as you answer the following questions about Millie's Famous Sandwich Spread.

   a. How many ounces of each ingredient are in a 4-ounce serving?
   b. How many grams of fat of each ingredient are in a 4-ounce serving?
   c. How many total grams of fat are in a 4-ounce serving?

Millie's Famous Sandwich Spread (4-ounce serving)

Ingredient	Ounces	Fat (g)
Chicken: Light Meat	____	____
Chicken: Dark Meat	____	____
Cream Cheese	____	____
Mayonnaise	____	____
Total Grams of Fat:		____

Use the bar graph and line graph for problem 2.

2. Fill in the table below as you answer these questions.

   a. What is Gregory Lin's oxygen usage rate (in liters per minute) for each of the listed activities? (Use Graph A.)
   b. About how many calories per minute does Gregory Lin use for each activity? (Knowing his oxygen usage rate, use Graph B to determine calorie use.)

Physiological Data Taken on Gregory Lin

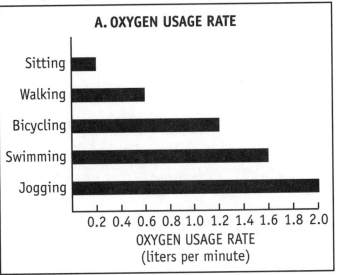

Activity	Oxygen Rate (Graph A)	Calorie Rate (Graph B)
Sitting	____	____
Walking	____	____
Bicycling	____	____
Swimming	____	____
Jogging	____	____

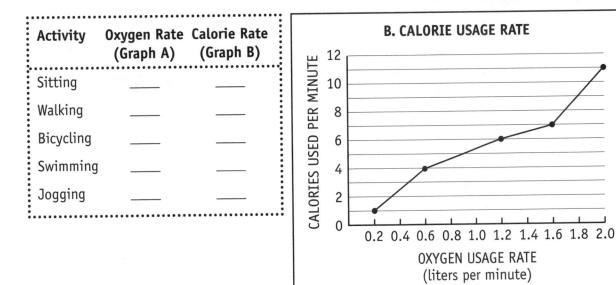

Use the following bar graph and table for problem 3.

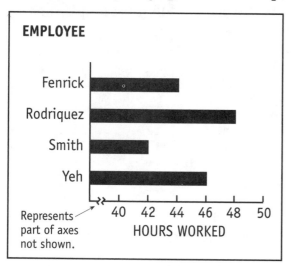

Hourly Pay Rate of Employees		
Employee	**Regular**	**Overtime**
Fenrick	$11.00	$16.50
Rodriquez	$11.00	$16.50
Smith	$11.00	$16.50
Yeh	$11.00	$16.50

3. Write your answers to the following questions in the table at the right.

   a. How much regular pay did each employee earn during the week shown?

   b. How much overtime pay did each employee earn during the week shown?

Employee	Regular Pay (first 40 hours)	Overtime Pay (hours over 40)
Fenrick		
Rodriquez		
Smith		
Yeh		

Use the picture graph and table below for problem 4.

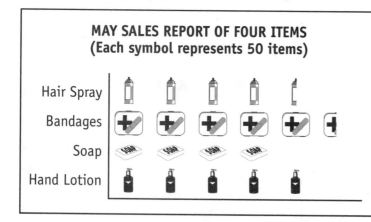

Family Market Price List		
Item	**Retail Price**	**Profit**
Hair Spray	$2.98	$1.05
Bandages	$1.99	$0.75
Soap	$0.89	$0.35
Hand Lotion	$5.77	$2.03

4. a. Determine the *approximate* total retail sales amount that Family Market received during May from the sale of bandages.

   b. Determine the *approximate* total profit that Family Market made during May from the sale of hand lotion.

Drawing Conclusions from Data

To **draw a conclusion** from data is to express an idea that is logically connected to the data. Tests often ask you to determine which of several conclusions is best supported by data given in a table or graph.

Median Earnings of Full-Time Workers

Year	Median Earnings Women	Median Earnings Men	Women's Earnings as a Percent of Men's Earnings
1960	$3,257	$5,368	60.7%
1970	$5,323	$8,966	59.4%
1975	$7,504	$12,758	58.8%
1980	$11,197	$18,612	60.2%
1985	$15,624	$24,195	64.6%
1989	$18,778	$27,430	68.0%

Source: Statistical Abstract of the United State: 1992

EXAMPLE   Which of the following statements is a conclusion that can be supported by data in the table?

a. Between 1960 and 1989, more women than men worked as secretaries and assistants, and these jobs paid less than the managerial jobs held mainly by men.

b. In 1989 the median salary of full-time working women was more than 30% less than the median salary of full-time working men.

c. Today most jobs pay men and women the same salary for the same work performed.

Statement a is true but *is not supported* by data in the table. The table does not deal with specific job categories.

Statement b is true and *is supported* by data in the table.

Statement c is false and *is not supported* by the data given. The table does not deal with earnings after 1989.

ANSWER: Only **statement b** is supported by data in the table.

(**Note:** Be aware of a statement that may be true but is not supported by the data given. You must carefully distinguish between conclusions that logically follow from given data and conclusions you know to be true from your own experience or from other sources of information.)

For problem 1, refer to the following bar graph.

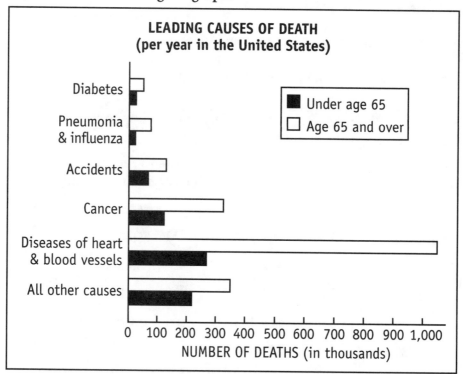

LEADING CAUSES OF DEATH
(per year in the United States)

Diabetes
Pneumonia & influenza
Accidents
Cancer
Diseases of heart & blood vessels
All other causes

■ Under age 65
□ Age 65 and over

0 100 200 300 400 500 600 700 800 900 1,000
NUMBER OF DEATHS (in thousands)

1. Which of the following statements is (are) supported by data on the bar graph? (More than one choice is possible.)

   a. A heart transplant is not an option in every case of severe heart disease.
   b. Less money is spent on diabetes research than for any other cause.
   c. For people over age 65, cancer is the leading cause of death.
   d. For people under age 65, heart and blood vessel disease is the leading cause of death.

For problem 2, refer to the circle graph at the right.

2. Which of the following statements is (are) supported by data on the circle graph? (More than one choice is possible.)

   a. The cost of obtaining energy from coal is about the same as obtaining energy from natural gas.
   b. The United States uses about twice as much energy from nuclear power as energy from hydroelectric power.
   c. The United States is the world's leading user of petroleum products.
   d. Less than half of the energy used in the United States comes from petroleum.

U.S. ENERGY CONSUMPTION
(1990)

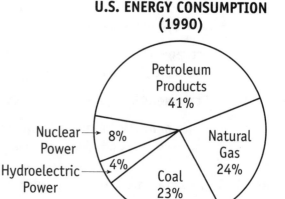

Petroleum Products 41%
Nuclear Power 8%
Hydroelectric Power 4%
Coal 23%
Natural Gas 24%

For problem 3, refer to the following bar graph and table.

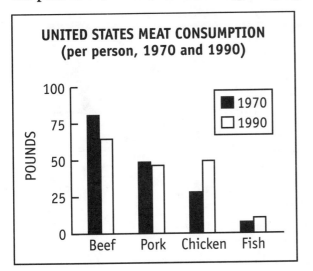

UNITED STATES MEAT CONSUMPTION
(per person, 1970 and 1990)

■ 1970
☐ 1990

POUNDS

Beef Pork Chicken Fish

**Nutritional Data for Selected Meats
(6-ounce serving)**

	Protein (g)	Fat (g)
Beef		
Hamburger	42	34
Steak	48	26
Pork		
Lean Pork	32	42
Ham	36	38
Chicken		
Dark Meat	48	11
Light Meat	54	9
Fish		
Salmon	34	10
Tuna	36	8

3. Which of the following statements is (are) supported by information given on the bar graph and table above? (More than one choice is possible.)

a. Between 1970 and 1990, Americans increased the amount of chicken and fish they ate but decreased the amount of beef and pork.

b. Between 1970 and 1990, the cost of beef and pork increased while the cost of chicken and fish decreased.

c. Between 1970 and 1990, Americans increased the amount of fat they obtained from eating the listed meats.

d. Between 1970 and 1990, Americans decreased the amount of fat they obtained from eating the listed meats.

For problem 4, refer to the following circle graphs.

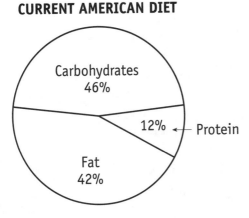

CURRENT AMERICAN DIET

Carbohydrates
46%

12% ← Protein

Fat
42%

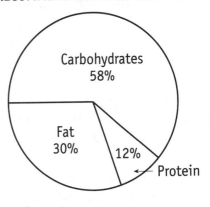

RECOMMENDED AMERICAN DIET

Carbohydrates
58%

Fat
30%

12%

← Protein

4. Which is the best summary of the recommendations shown?

a. Eat fewer carbohydrates and more fat.

b. Eat less protein and more carbohydrates.

c. Eat less fat and more protein.

d. Eat less fat and more carbohydrates.

Understanding Probability

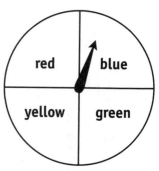

Probability is the study of **chance**—the likelihood of an event happening. The word *chance* indicates our lack of control over what actually happens.

The spinner circle at the right is divided into four equal-size sections: red, blue, green, and yellow. Assuming the spinner doesn't stop on a line, it is equally likely to stop in any of these four sections. We say, "Where the spinner stops is left to chance."

In the study of probability, the outcome (result) we're interested in is called a **favorable outcome**. The **probability of a favorable outcome** is defined as follows.

probability of a favorable outcome $= \dfrac{\text{number of favorable outcomes}}{\text{total number of possible outcomes}}$

The **number of favorable outcomes** is simply the number of ways that the favorable outcome can occur.

Expressing Probability as a Number

Probabilities are expressed as numbers (usually common fractions) ranging from 0 to 1 or as percents from 0% to 100%.

EXAMPLE 1   What is the probability that the spinner pictured above will stop on green?

> **STEP 1**   Notice there are 4 *possible outcomes,* but only 1 *favorable outcome*—only 1 green section.

> **STEP 2**   Write the probability as a fraction. $\dfrac{\text{favorable outcomes}}{\text{possible outcomes}} = \dfrac{1}{4}$ (one in four)

ANSWER: The probability of a green outcome is $\frac{1}{4}$, **or 25%.** On the average, only **1 spin in 4** will stop on green.

A *probability of 0 (0%)* means that an event cannot occur. The probability is 0 that the spinner will stop on brown because there is no brown section!

A *probability of 1 (100%)* means that an outcome will definitely occur. The probability is 1 that the spinner will stop somewhere in the circle. There is no other possibility, assuming the spinner can't keep spinning forever!

The value of a probability is always a number from 0 to 1. It is never less than 0 and it is never greater than 1.

- A probability smaller than $\frac{1}{2}$ means that an event will happen less than half of the time—the smaller the probability, the less likely that the event will happen.
- A probability larger than $\frac{1}{2}$ means that an event will happen more than half of the time—the larger the probability, the more likely that the event will happen.

EXAMPLE 2 Look at the spinner at the right. What is the probability that this spinner will stop on a section labeled $10?

STEP 1 Notice there are *6 possible outcomes.* Of these 6 outcomes, 2 are labeled $10. Thus, the number of *favorable outcomes* is 2.

STEP 2 Write the probability as a fraction. Reduce this fraction.

$$\frac{\text{favorable outcomes}}{\text{possible outcomes}} = \frac{2}{6} = \frac{1}{3} \text{ (one in three)}$$

On the average, 1 spin in 3 will stop on a $10 section.

ANSWER: The probability of a $10 outcome is $\frac{1}{3}$, **or $33\frac{1}{3}$%.**

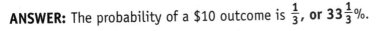

Solve each problem below.

1. There are six faces on a cube. Each side is equally likely to be up after the cube is tossed.

   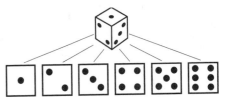

   a. What is the probability of rolling a 4 with one toss of the cube?

   b. What is the probability of rolling an even number (2, 4, or 6) with one toss of the cube?

2. At the right is a spinner for a board game.

   a. What is the probability that a player will win $25 on one spin?

   b. What is the probability that a player will win $50 or more on one spin?

3. A box contains 4 pennies, 12 nickels, and 20 dimes. Suppose you reach into the box and randomly take out 1 coin. What is the probability that the coin you choose will be

   a. a penny?   **b.** a nickel?   **c.** a dime?   **d.** a quarter?

4. If you randomly pick a number from 1 to 30, what is the probability that the number you pick will be evenly divisible by 4?

5. Suppose you shut your eyes and choose a penny from those shown.

   a. What is the probability that you will choose the heads-up penny?

   b. What is the probability that you will *not* choose the heads-up penny?

Problem Solver: Making a List to Predict Outcomes

For many probability problems, making a list of possible outcomes is a good first step. Writing information in this organized way makes it much easier for you to count the number of favorable outcomes.

EXAMPLE When you roll a pair of cubes, what is the probability that the sum of the two cubes will be 7? Asked another way, "What is the probability that you will roll a 7?"

In this problem, each outcome consists of the sum of the two numbers shown face up on the rolled cubes. First write a list of all combinations of pairs of numbers that the cubes can have.

STEP 1 Make a list of all possible outcomes. Notice that for each number that cube #1 can have, there are six values for cube #2. As you see, there are 36 possible outcomes.

Cube #1	Cube #2	Cube #1	Cube #2	Cube #1	Cube #2
1	1	3	1	5	1
1	2	3	2	5	2
1	3	3	3	5	3
1	4	3	4	5	4
1	5	3	5	5	5
1	6	3	6	5	6
2	1	4	1	6	1
2	2	4	2	6	2
2	3	4	3	6	3
2	4	4	4	6	4
2	5	4	5	6	5
2	6	4	6	6	6

STEP 2 Count the number of favorable outcomes—the number of ways a sum of 7 can occur. In the list above, a box is drawn around each of these. As shown, there are six favorable outcomes.

STEP 3 Write the probability as a fraction.

$$\frac{\text{favorable outcomes}}{\text{total outcomes}} = \frac{6}{36} = \frac{1}{6}$$

ANSWER: The probability of rolling a 7 is $\frac{1}{6}$.

For comparison, notice that there is only one way to roll a 2—each cube coming up 1. Thus, you are six times as likely to roll a 7 as you are to roll a 2.

Solve each problem below.

1. Look at the list of outcomes in the example on page 168. When you roll a pair of cubes, what is the probability that you will roll

    a. an 8?
    b. an 11?

2. Amy, Gail, and Vicki are going to a play. They have reserved seats #26, #27, and #28 in the 4th row.

    a. Complete the list at the right to determine how many different ways the girls can be seated.
    b. If each girl randomly chooses a ticket, what is the probability that Amy will sit next to Vicki?

#26	#27	#28
Amy	Gail	Vicki
Amy	Vicki	Gail
Gail		

3. At his restaurant, Julian serves orange, cola, and root beer soft drinks. He also serves hamburgers, fish sandwiches, and chicken sandwiches.

    a. How many soft drink-sandwich combinations are possible at Julian's?
    b. If each combination is equally likely to be ordered, what is the probability that Julian's next customer will order a root beer and a fish sandwich?

4. Look again at the list of outcomes in the example on page 168.

    a. When you roll a pair of cubes, how many more times likely are you to roll a 7 than a 4?
    b. When you roll a pair of cubes, what is the probability you will roll "doubles" (two identical numbers)?

5. Phil is going to place three posters side-by-side across the store window. The posters advertise styles in ties, belts, and shirts.

    a. Write a list at the right to show the number of different ways Phil can arrange these posters.
    b. If Phil randomly chooses how to arrange the posters, what's the probability that the tie poster will *not* be next to the shirt poster?

1st	2nd	3rd

6. The first three letters of a license plate were reported to be G, N, and B, but the caller couldn't remember the exact order.

    a. How many different orderings of these three letters are possible?
    b. What is the probability that the correct number starts with N?
    c. What is the probability that the correct number starts with either a G *or* a B?

Finding Probability for Two Events

Successive events are events that happen one after another. To find the probability of successive events, multiply the probability of the first event by the probability of the second event.

Independent Events

> The probability of a second event does not depend on the first event.

EXAMPLE 1 When rolling a cube, what is the probability of rolling two 6s in a row?

STEP 1 The probability of rolling a 6 on either roll is $\frac{1}{6}$.

STEP 2 Multiply the first probability ($\frac{1}{6}$) by the second ($\frac{1}{6}$).

$$\frac{1}{6} \times \frac{1}{6} = \frac{1}{36}$$

ANSWER: The probability of rolling two 6s is $\frac{1}{36}$.

When you roll a cube twice in a row, rolling

followed by

happens only once every 36 times.

Dependent Events

> The probability of the second event depends on the first event.

EXAMPLE 2 If you randomly choose 2 cards from the 5 cards shown below, what is the probability that you will choose 2 face cards?

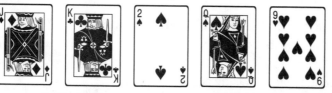

STEP 1 Notice that for your first choice there are 5 possible outcomes, 3 of which are face cards. So the probability of drawing a face card on your first draw is $\frac{3}{5}$.

STEP 2 Now assume you *do* draw a face card on your first draw. This means that there are 4 possible outcomes for your second draw, only 2 of which are face cards. The probability of drawing a face card on your second draw is then $\frac{2}{4} = \frac{1}{2}$.

STEP 3 Multiply the first probability by the second.
$$\frac{3}{5} \times \frac{1}{2} = \frac{3}{10}$$

ANSWER: The probability of choosing 2 face cards is $\frac{3}{10}$. When you draw two cards in a row from the face cards shown, you'll draw 2 face cards **3 out of every 10 times.**

Solve each problem below. Determine the probability of both the first and second events before multiplying.

1. When you are rolling a cube, what is the probability of rolling a 5 and then rolling the cube again and getting a 2?

2. The median high temperature in June in Salem is 82°F. Estimate the probability that on 2 days in a row in June the temperature will rise above 82°F. Assume that the daily high temperature is equally likely to be higher than 82°F or lower than 82°F.

3. A coin has two sides: heads and tails.

   a. If you flip a coin twice in a row, what is the probability of getting 2 heads?
   b. If you flip a coin three times in a row, what is the probability of getting 3 heads?

4. Suppose you randomly choose two cards from the group of cards shown.

   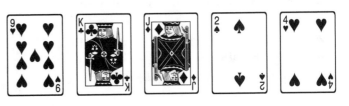

   a. What is the probability that the first card you choose will be a face card?
   b. What is the probability that both cards will be face cards?

5. Seven friends are visiting seven senior citizens at a retirement home. The names of the seven senior citizens, 4 women and 3 men, are written on slips of paper. Each friend randomly chooses the name of a senior citizen. What is the probability that the first two people who draw will draw a man's name?

6. Audrey's son has a bag of 8 blue marbles and 4 green marbles. He tells Audrey to close her eyes and choose 2. What is the probability that she will select one marble of each color? (**Hint:** Assume that Audrey gets either a blue or green marble on her first draw.)

7. In Bill's pocket are 5 dimes, 3 nickels, and 2 pennies. If Bill randomly takes 2 coins out of his pocket, what is the probability that he will take out 15¢?

8. With your eyes closed, suppose you choose 3 pennies from the group.

   a. What is the probability that the first 2 pennies you choose will be heads up?
   b. What is the probability that all 3 pennies will be heads up?

Problem Solver: Using Probability for Prediction

Probability is often used as a basis for making predictions. To predict the number of times an event will occur out of a large number of tries, you multiply the probability of the single event by the number of tries.

EXAMPLE 1 If you roll a cube 300 times, about how many 6s will you roll?

STEP 1 Determine the probability of rolling a 6 on one roll.

$$\frac{\text{favorable outcomes}}{\text{total outcomes}} = \frac{1}{6}$$

STEP 2 Multiply 300 by $\frac{1}{6}$.

$$300 \times \frac{1}{6} = \textbf{50}$$

ANSWER: It is likely that you will roll **50** 6s.

(**Note:** If each of 100 people rolls a cube 300 times, it is possible that no one will roll exactly 50 6s. But the average would be about 50.)

EXAMPLE 2 If you roll a pair of cubes 300 times, about how many double 6s will you roll?

STEP 1 From the example on page 170, the probability of rolling 2 6s on 1 roll is $\frac{1}{36}$.

STEP 2 Multiply 300 by $\frac{1}{36}$.

$$399 \times \frac{1}{36} = \textbf{8}\frac{1}{3}$$

ANSWER: It is likely that you will roll **8** double 6s.

(**Note:** Example 2 requires a whole-number answer. In probability problems such as this one, round your result to the nearest whole number.)

Basing Probability on an Outcome Pattern

Up to now, you've studied how probability can be based on the laws of chance. Now, you'll learn how probability can also be based on an **outcome pattern**—a pattern formed by a large number of previous outcomes. The pattern gives the probability of a similar future outcome.

EXAMPLE 3 Clarese, a basketball player, has made 70% of her free throws so far this season. What is the probability that Clarese will make her first two free throws tonight?

STEP 1 Due to Clarese's past performance, you say that the probability she will hit any 1 free throw is 70% (or $\frac{7}{10}$).

STEP 2 The probability that she will make 2 in a row is found by multiplying probabilities.

$$\frac{7}{10} \times \frac{7}{10} = \frac{49}{100} \leftarrow \text{probability of making both free throws}$$

probability of making ⟶ first free throw — probability of making second free throw

ANSWER: $\frac{49}{100}$ **or about 50%** (About half the time, a 70% shooter will make 2 in a row!)

Solve each problem.

1. If you roll a cube 500 times, about how many 3s will you roll?

2. On a production line, 9 out of the last 200 microchips were defective.

   a. What is the probability that the next microchip will be defective?
   b. Out of the next 750 microchips, what is the most likely number that will be defective?

3. a. What is the probability that the next car Garner Motors sells will be a foreign car (not domestic)?
   b. If Garner Motors sells 75 cars next month, about how many will be foreign cars?

GARNER MOTORS SALES

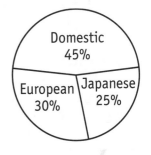

4. A survey of 1,000 people had these results: 180 preferred bacon and eggs for breakfast, 250 preferred pancakes, 380 preferred cereal, and 190 had other choices.

   a. Based on this survey, what is the probability that the next person asked will prefer pancakes for breakfast?
   b. Of the next 200 people surveyed, about how many will say they prefer cereal for breakfast?

5. Georgia, a softball pitcher, has struck out 10% of the batters she's faced this season.

   a. What is the probability that Georgia will strike out the next batter she faces?
   b. What is the probability that Georgia will strike out the next two batters she faces?

6. The weather bureau reports that there is a 40% chance for rain tomorrow in Bend and a 25% chance for rain in Newport. The same forecast has already been made 30 times this year!

   a. If the forecast is correct, what is the probability that it will rain in both Bend *and* Newport tomorrow?
   b. About how many times this year did it rain in both cities the day after this forecast?

7. What is the probability that the next two projects undertaken by WIX Co. will be completed early?

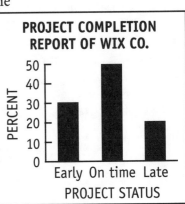

Data Analysis and Probability Review

Solve the problems below.

1. What was the approximate median family income in the U.S. in 1975?

2. What was the approximate median family income in the U.S. in 1983?

3. If the trend shown in the graph continues, what will be the approximate median family income in the U.S. in the year 2005?

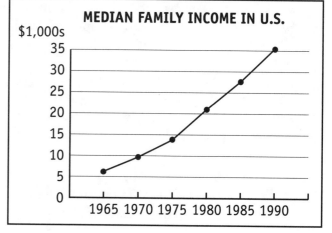

4. Which conclusion is best supported by data in the bar graph?

   a. Radio was invented before TV.
   b. The cost of cable TV is on the rise.
   c. More families have a radio than a TV.
   d. The cost of VCRs is decreasing.
   e. VCRs are more popular than cable TV.

5. Which conclusion is *not* supported by data in the bar graph?

   a. The cost of watching movies on a VCR is less than watching them on cable TV.
   b. Between 1980 and 1990, the percent of American households with a radio changed little, if any.
   c. By 1990, cable TV was in more than half of all American households.
   d. In 1990, more American households had a radio than a telephone.
   e. In 1990, more American households had a VCR than cable TV.

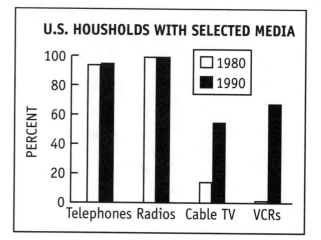

6. You can reasonably predict that by the year 2010, the percent of American households with VCRs will be

   a. about 10%   b. about 30%   c. about 50%   d. about 70%   e. 90% or more

7. Joey bought a bag of popsicles. Of the popsicles, 3 are orange, 5 are grape, and the remaining 4 are cherry. In the car, Joey's daughter reaches into the bag and takes a popsicle without looking. What is the probability that the popsicle she takes will be cherry?

8. Tim has 3 shirts; 1 blue, 1 white, and 1 brown. He also has 3 ties: 1 striped, 1 plain, and 1 flowered. If Tim randomly chooses a shirt and tie, what is the probability that he'll wear the white shirt with the striped tie?

9. Suppose you randomly choose two quarters from the group at the right. What is the probability that both quarters will be heads up?

   a. 10%   **d.** 30%
   b. 20%   **e.** 50%
   c. 25%

10. Following a TV call-in show, 240 viewers phoned the station. Of these, 160 prefer a 1-hour news show, 60 prefer a half-hour news show, and 20 have no preference. Of the next 100 callers, about how many will prefer the half-hour news show?

11. On the first day of registration, 25 boys and 25 girls signed up for swimming lessons. If that trend continues, what is the probability that the next two people who register will both be girls?

12. If the trend shown in the graph continues, what is the probability that the next two students who register at City College will be over 35 years of age?

    a. $\frac{1}{25}$ **c.** $\frac{1}{5}$ **e.** $\frac{7}{10}$

    b. $\frac{1}{10}$ **d.** $\frac{2}{5}$

AGE OF STUDENTS AT CITY COLLEGE

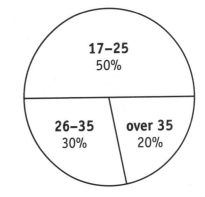

ALGEBRA

Solving Addition and Subtraction Equations

Algebra is a powerful problem-solving tool that is the basic mathematical language of all technical fields, from auto mechanics to nursing to electronics. In algebra, you use a letter, called a **variable,** to stand for each unknown quantity you're trying to find.

Here are examples of four basic algebra equations with variables.

Basic Algebra Equations	Solutions	Check
Addition equation: $x + 4 = 9$	$x = 5$	$\mathbf{5} + 4 = 9$
Subtraction equation: $y - 3 = 12$	$y = 15$	$\mathbf{15} - 3 = 12$
Multiplication equation: $2z = 14$	$z = 7$	$2(\mathbf{7}) = 14$
Division equation: $\frac{n}{7} = 5$	$n = 35$	$\frac{\mathbf{35}}{7} = 5$

Finding the value of the variable that makes the equation a true statement is called **solving the equation.** The correct value of the variable is called the **solution.**

To **check a solution,** substitute the solution for the unknown in the original equation.

To solve an addition equation, subtract the added number from each side of the equation so that the variable is alone.

> To simplify an equation, drop expressions such as $8 - 8$ that equal 0.

EXAMPLE 1 Solve: $x + 8 = 14$

$$x + 8 = 14$$

    STEP 1 Subtract 8 from each side of the equation.

$$x + 8 - 8 = 14 - 8$$

    STEP 2 Simplify each side. ($x + 0 = x$; $14 - 8 = 6$)

$$x = 6$$

ANSWER: $x = 6$

Check: $6 + 8 = 14$
              ✓$14 = 14$

To solve a subtraction equation, add the subtracted number to each side of the equation.

EXAMPLE 2 Solve: $y - 7 = 9$

$$y - 7 = 9$$

    STEP 1 Add 7 to each side of the equation.

$$y - 7 + 7 = 9 + 7$$

    STEP 2 Simplify each side. ($y + 0 = y$; $9 + 7 = 16$)

$$y = 16$$

ANSWER: $y = 16$

Check: $16 - 7 = 9$
              ✓$9 = 9$

Check to see if the value given for the variable is the solution to each equation. Circle Yes if the given value is the solution and No if it is not.

1. $x + 7 = 9$ Try $x = 3$.   Yes   No

2. $y - 8 = 24$ Try $y = 16$.   Yes   No

3. $n + 5 = 13$ Try $n = 8$.   Yes   No

4. $a - 4 = 8$ Try $a = 12$.   Yes   No

Solve each addition equation by subtracting the added number from each side of the equation. Check each answer. The first problem in each row is partially completed.

5. $x + 4 = 12$ $y + 5 = 9$ $n + 2 = 5$ $m + 8 = 16$
    $x + 4 - 4 = 12 - 4$

6. $y + 9 = 15$ $z + 7 = 19$ $x + 12 = 30$ $n + 3 = 13$
    $y + 9 - 9 = 15 - 9$

7. $z + \frac{3}{2} = 4$ $x + \frac{2}{3} = 2$ $y + \frac{7}{5} = \frac{11}{5}$ $n + 1\frac{3}{4} = 3\frac{2}{3}$

    $z + \frac{3}{2} - \frac{3}{2} = 4 - \frac{3}{2}$

Solve each subtraction equation by adding the subtracted number from each side of the equation. Check each answer.

8. $x - 8 = 7$ $y - 5 = 12$ $z - 3 = 7$ $n - 2 = 0$
    $x - 8 + 8 = 7 + 8$

9. $m - 7 = 13$ $x - 1 = 17$ $y - 14 = 3$ $p - 4 = 4$
    $m - 7 + 7 = 13 + 7$

10. $x - \frac{5}{8} = 3$ $y - \frac{1}{3} = 2$ $z - \frac{5}{4} = \frac{15}{8}$ $n - 2\frac{3}{4} = 3\frac{1}{8}$

    $x - \frac{5}{8} + \frac{5}{8} = 3 + \frac{5}{8}$

Solving Multiplication and Division Equations

In the equation $6x = 12$, the number 6 is called the **coefficient** of the variable x. To solve a multiplication equation, divide each side of the equation by the coefficient of the variable.

EXAMPLE 1   Solve: $6x = 42$ $6x = 42$

STEP 1   Divide each side of the equation $\frac{6x}{6} = \frac{42}{6}$ (The 6s on the left cancel
by 6.   because $\frac{6}{6} = 1$ and $1x = x$.)

STEP 2   Simplify each side. $x = 7$

ANSWER: $x = 7$

Check: $6(7) = 42$
✓$42 = 42$

To solve a division equation, multiply each side of the equation by the number that divides the variable.

EXAMPLE 2   Solve: $\frac{n}{4} = 9$ $\frac{n}{4} = 9$

STEP 1   Multiply each side of the equation $\frac{n}{4}(4) = 9(4)$ (The 4s on the left cancel
by 4.   because $\frac{1}{4} \times 4 = \frac{1}{4} \times \frac{4}{1} = \frac{4}{4} = 1$.)

STEP 2   Simplify each side. $n = 36$

ANSWER: $n = 36$

Check: $\frac{36}{4} = 9$
✓$9 = 9$

The **reciprocal** of a fraction is the new fraction you get by switching the numerator and denominator. For example, the reciprocal of $\frac{2}{3}$ is $\frac{3}{2}$ and the reciprocal of $\frac{1}{7}$ is $\frac{7}{1}$ or 7. If a fraction multiplies an unknown, multiply each side of the equation by the reciprocal of the fraction.

EXAMPLE 3   Solve: $\frac{2}{3}x = 10$ $\frac{2}{3}x = 10$

STEP 1   Multiply each side of the equation $\frac{3}{2}(\frac{2}{3})x = \frac{3}{2}(10)$ (On the left, $\frac{3}{2}$ cancels $\frac{2}{3}$
by the reciprocal of $\frac{2}{3}$, which is $\frac{3}{2}$.   because $\frac{3}{2} \times \frac{2}{3} = \frac{6}{6} = 1$.)

STEP 2   Simplify each side. $x = 15$

ANSWER: $x = 15$ Check: $\frac{2}{3}(15) = 10$
✓$10 = 10$

Check to see if the value given for the variable is the solution to each equation. Circle Yes if the given value is the solution and No if it is not.

1. $7x = 63$ Try $x = 8$.   Yes   No

2. $\frac{y}{4} = 12$ Try $y = 48$.   Yes   No

3. $\frac{2}{3}z = 12$ Try $z = 21$.   Yes   No

Solve each multiplication equation by *dividing each side of the equation by the number that multiplies the variable*. Check each answer. The first problem in each row is partially completed.

4. $2y = 14$ $3z = 24$ $5n = 45$ $7z = 42$
   $\frac{2y}{2} = \frac{14}{2}$

5. $2x = 3.5$ $4y = 10.4$ $3z = \frac{13}{2}$ $5n = 3\frac{1}{2}$
   $\frac{2x}{2} = \frac{3.5}{2}$

Solve each division equation by *multiplying each side of the equation by the number that divides the variable*. Check each answer.

6. $\frac{x}{3} = 12$ $\frac{y}{5} = 9$ $\frac{z}{7} = 5$ $\frac{n}{6} = 12$
   $\frac{x}{3}(3) = 12(3)$

7. $\frac{n}{2} = 2\frac{1}{4}$ $\frac{x}{3} = 4.5$ $\frac{y}{4} = 3\frac{3}{5}$ $\frac{p}{7} = 2.8$

Solve each equation by *multiplying each side of the equation by the reciprocal of the fraction that multiplies the variable*. Check each answer.

8. $\frac{2}{5}x = 14$ $\frac{3}{4}y = 12$ $\frac{7}{8}z = 14$ $\frac{11}{16}n = 33$
   $\frac{5}{2}\left(\frac{2}{5}\right)x = \frac{5}{2}(14)$

Solving Multistep Equations

A **multistep equation** contains two or more of the operations of addition, subtraction, multiplication, and division.

To solve a multistep equation *containing variables,* follow these two rules.

1. Do addition or subtraction first.

2. Do multiplication or division last.

EXAMPLE 1 Solve: $3y - 9 = 36$

$$3y - 9 = 36$$

STEP 1 Add 9 to each side of the equation. $3y - 9 + 9 = 36 + 9$

Simplify each side. $3y = 45$

STEP 2 Divide each side of the equation by 3. $\frac{3y}{3} = \frac{45}{3}$

Simplify each side. $y = 15$

ANSWER: $y = 15$

Check: $3(15) - 9 = 36$
$45 - 9 = 36$
$\checkmark 36 = 36$

EXAMPLE 2 Solve: $\frac{x}{5} + 6 = 8$

$$\frac{x}{5} + 6 = 8$$

STEP 1 Subtract 6 from each side of the equation. $\frac{x}{5} + 6 - 6 = 8 - 6$

Simplify each side. $\frac{x}{5} = 2$

STEP 2 Multiply each side of the equation by 5. $\frac{x}{5}(5) = 2(5)$

Simplify each side. $x = 10$

ANSWER: $x = 10$

Check: $\frac{10}{5} + 6 = 8$

$2 + 6 = 8$
$\checkmark 8 = 8$

Solve each equation. (Remember: At each step, you must perform the same operation on each side of the equation.)

1. $2x + 8 = 10$ $\qquad$ $3y - 4 = 8$ $\qquad$ $4n + 7 = 11$ $\qquad$ $5z - 9 = 36$

2. $8m + 11 = 43$ $\qquad$ $4p - 9 = 27$ $\qquad$ $7y + 3.5 = 17.5$ $\qquad$ $5x = 12\frac{1}{2}$

3. $\frac{x}{5} + 4 = 7$ $\qquad$ $\frac{y}{3} - 7 = 3$ $\qquad$ $\frac{z}{6} + 8 = 9$ $\qquad$ $\frac{n}{4} - 8 = 12$

4. $\frac{z}{3} - 1 = 6.5$ $\qquad$ $\frac{m}{4} + \frac{3}{2} = \frac{8}{2}$ $\qquad$ $\frac{y}{5} - 3\frac{1}{3} = 2\frac{2}{3}$ $\qquad$ $\frac{p}{6} + 8 = 14\frac{1}{4}$

5. $\frac{3}{4}x + 7 = 16$ $\qquad$ $\frac{2}{3}y - 9 = 9$ $\qquad$ $\frac{4}{5}z + 12 = 20$ $\qquad$ $\frac{8}{5}n - 4 = 20$

6. $\frac{7}{4}x - 9 = 26$ $\qquad$ $\frac{4}{3}z + \frac{5}{3} = \frac{14}{3}$ $\qquad$ $2\frac{1}{2}x + 4 = 19$ $\qquad$ $4\frac{2}{3}y - 7 = 21$

$\qquad\qquad\qquad\qquad\qquad\qquad\qquad\qquad\qquad$ (**Hint:** Write $2\frac{1}{2}$ as $\frac{5}{2}$.)

Applying Your Skills

Learning to use equations to solve word problems is a very important part of your study of algebra. On these next two pages, you'll learn to write and solve equations for problems that you've already learned to solve—without using algebra! See if algebra simplifies these problems.

To solve a word problem using algebra, follow these steps.

STEP 1 Represent the unknown with a letter (usually the letter x).

STEP 2 Write an equation that represents the given information.

STEP 3 Solve the equation to find the unknown value.

__EXAMPLE 1__ Five times a number is equal to 75. What is the number?

STEP 1 Let x = unknown number.

STEP 2 Write an equation for x. $\qquad\qquad\qquad\qquad$ $5x = 75$

STEP 3 Solve the equation.

Divide each side by 5; then simplify. $\qquad\qquad$ $\frac{5x}{5} = \frac{75}{5}$

ANSWER: $x = 15$ $\qquad\qquad\qquad\qquad\qquad\qquad\qquad\qquad$ $x = 15$

$$\text{Check:} \qquad 5(15) = 75$$
$$\checkmark 75 = 75$$

__EXAMPLE 2__ Muriel earns \$36 each day in salary as a waitress. She also earns \$2.50 per hour as her share of tips. On Tuesday, Muriel earned a total of \$51. How many hours did Muriel work on Tuesday?

STEP 1 Let x = number of hours Muriel worked on Tuesday.

STEP 2 Write an equation for x. $\qquad\qquad$ $\$2.50x + \$36 = \$51$

STEP 3 Solve the equation.

a. Subtract \$36 from each side; $\qquad$ $\$2.50x + \$36 - \$36 = \$51 - \$36$

then simplify. $\qquad\qquad\qquad\qquad\qquad$ $\$2.50x = \$15$

b. Divide each side by \$2.50; $\qquad\qquad$ $\frac{\$2.50x}{\$2.50} = \frac{\$15}{\$2.50}$

then simplify. $\qquad\qquad\qquad\qquad\qquad\qquad$ $x = 6$

ANSWER: $x = 6$ hours

$$\text{Check:} \quad \$2.50(6) + \$36 = \$51$$
$$\$15 + \$36 = \$51$$
$$\checkmark \$51 = \$51$$

In each problem below, represent the unknown value as a variable, write an equation, and solve the equation.

1. One fourth of Kari's salary goes for her share of rent. If Kari's monthly rent payment is $375, what is Kari's monthly salary?

    a. Variable: x = Kari's monthly salary

    b. Equation:

    c. Solution:

2. Joseph paid $320 for a new color TV. If Joseph was given a $65 discount, what was the original price of the set?

    a. Variable: p = original price of set

    b. Equation:

    c. Solution:

3. For 8 hours of yard work, Vince was paid $71. Of this amount, $15 was a bonus. What was Vince's regular hourly rate?

    a. Variable: r = regular hourly rate

    b. Equation:

    c. Solution:

4. Wendy has a part-time job. Each month, Wendy deposited $\frac{1}{10}$ of her check in a savings account. After six deposits, she placed an additional $150 in this account. If she now has $900 in savings, how much does Wendy usually deposit each month?

    a. Variable: d =

    b. Equation:

    c. Solution:

5. At Hoover Elementary School, 60% of the students are in the grades kindergarten through third grade. If 180 children are in these four grades, what is the total enrollment at Hoover Elementary?

    a. Variable:

    b. Equation:

    c. Solution:

6. Jackie cut a ribbon into eight equal pieces. She measured and found that each piece was 2 inches longer than a foot-long ruler. What was the length of the original uncut ribbon?

    a. Variable:

    b. Equation:

    c. Solution:

Combining Like Terms in an Equation

··

An equation is made up of **terms.** Each term is a number standing alone or is a variable (representing an unknown value) multiplied by its coefficient.

When the same variable appears in more than one term in an addition or subtraction equation, you combine the separate *like terms* by adding or subtracting the coefficients. The following examples show how separate variables are combined.

$4x + 3x = 7x$ since $4 + 3 = 7$

$8y - 2y = 6y$ since $8 - 2 = 6$

$\frac{3}{2}n + n = \frac{5}{2}n$ since $\frac{3}{2} + 1 = \frac{3}{2} + \frac{2}{2} = \frac{5}{2}$

$9z - z = 8z$ since $9 - 1 = 8$

(**Note:** When a variable stands alone, the coefficient is understood to be 1. Therefore, in the examples shown above, the coefficient of n is 1 and the coefficient of z is 1.)

<u>EXAMPLE 1</u> Solve: $7x + 3x = 100$ $7x + 3x = 100$

     STEP 1 Combine like terms. $10x = 100$

     STEP 2 Divide each side by 10. $\frac{10x}{10} = \frac{100}{10}$

             Simplify each side. $x = \mathbf{10}$

ANSWER: $x = \mathbf{10}$

                                  Check: $7(\mathbf{10}) + 3(\mathbf{10}) = 100$

                                            $70 + 30 = 100$

                                            $\checkmark 100 = 100$

<u>EXAMPLE 2</u> Solve: $12n - 4n + 10 = 42$ $12n - 4n + 10 = 42$

     STEP 1 Combine like terms. $8n + 10 = 42$

     STEP 2 Subtract 10 from each side. $8n + 10 - 10 = 42 - 10$

             Simplify each side. $8n = 32$

     STEP 3 Divide each side by 8. $\frac{8n}{8} = \frac{32}{8}$

             Simplify each side. $n = \mathbf{4}$

ANSWER: $n = \mathbf{4}$

                         Check: $12(\mathbf{4}) - 4(\mathbf{4}) + 10 = 42$

                                     $48 - 16 + 10 = 42$

                                        $32 + 10 = 42$

                                           $\checkmark 42 = 42$

Combine each pair of like terms.

1. $3x + 2x$ $\qquad$ $7n - 4n$ $\qquad$ $9y + y$ $\qquad$ $12z - 2z$

2. $\frac{1}{2}x + \frac{1}{4}x$ $\qquad$ $\frac{11}{8}y - y$ $\qquad$ $\frac{2}{3}n + \frac{1}{2}n$ $\qquad$ $1\frac{5}{8}w - \frac{6}{16}w$

Solve each equation.

3. $6x + 2x = 32$ $\qquad$ $5y - y = 12$ $\qquad$ $4z + 9z = 39$ $\qquad$ $9m - 4m = 45$

4. $\frac{2}{3}x + \frac{2}{3}x = 3$ $\qquad$ $\frac{7}{8}n - \frac{3}{8}n = 2$ $\qquad$ $\frac{7}{5}z - \frac{3}{10}z = 4$ $\qquad$ $\frac{3}{4}x - \frac{2}{3}x = 1$

5. $1.4z + 0.6z = 17$ $\qquad$ $2.6x - 0.7x = 38$ $\qquad$ $6y - 3.75y = 9$ $\qquad$ $4.5x + 3x = 61.5$

6. $9n - 4n + 18 = 28$ $\qquad$ $2z - z + 4 = 11$ $\qquad$ $5y + 2y - 5 = 7$ $\qquad$ $12x + 3x + 14 = 76$

7. $3x + x - 9 = 47$ $\qquad$ $8q - 3q + 15 = 40$ $\qquad$ $m + 3m + 6 = 18$ $\qquad$ $5n + 3n - 12 = 8$

Solving Equations with Terms on Both Sides

A variable (representing an unknown value) can appear in terms on both sides of an equation. Usually, you'll move all terms containing a variable to the left side of the equation.

EXAMPLE 1 Solve: $5x = 3x + 12$

$$5x = 3x + 12$$

STEP 1 Subtract $3x$ from each side.

$$5x - 3x = 3x - 3x + 12$$

Simplify each side.

$$2x = 12$$

STEP 2 Divide each side by 2.

$$\frac{2x}{2} = \frac{12}{2}$$

Simplify each side.

$$x = 6$$

ANSWER: $x = 6$

Check: $5(\mathbf{6}) = 3(\mathbf{6}) + 12$
$30 = 18 + 12$
$\checkmark 30 = 30$

EXAMPLE 2 Solve: $6y - 14 = 2y + 82$

$$6y - 14 = 2y + 82$$

STEP 1 Subtract $2y$ from each side.

$$6y - 2y - 14 = 2y - 2y + 82$$

Simplify each side.

$$4y - 14 = 82$$

STEP 2 Add 14 to each side.

$$4y - 14 + 14 = 82 + 14$$

Simplify each side.

$$4y = 96$$

STEP 3 Divide each side by 4.

$$\frac{4y}{4} = \frac{96}{4}$$

Simplify each side.

$$y = 24$$

ANSWER: $y = 24$

Check: $6(\mathbf{24}) - 14 = 2(\mathbf{24}) + 82$
$144 - 14 = 48 + 82$
$\checkmark 130 = 130$

Complete each problem. Check each answer.

1.
$6x + 9 = 2x + 25$
$6x - 2x + 9 = 2x - 2x + 25$
$4x + 9 = 25$
$4x + 9 - 9 = 25 - 9$

$6z - 19 = 3z + 47$
$6z - 3z - 19 = 3z - 3z + 47$
$3z - 19 = 47$
$3z - 19 + 19 = 47 + 19$

$11n - 13 = 4n + 36$
$11n - 4n - 13 = 4n - 4n + 36$
$7n - 13 = 36$
$7n - 13 + 13 = 36 + 13$

Check:
$6(\ ) + 9 = 2(\ ) + 25$

Check:
$6(\ ) - 19 = 3(\ ) + 47$

Check:
$11(\ ) - 13 = 4(\ ) + 36$

2.

$5y - 7 = 3y + 23$	$4x + 12 = x + 39$	$13p + 28 = 5p + 52$
$5y - 3y - 7 = 3y - 3y + 23$	$4x - x + 12 = x - x + 39$	$13p - 5p + 28 = 5p - 5p + 52$
$2y - 7 = 23$	$3x + 12 = 39$	$8p + 28 = 52$

Check:

$5(\quad) - 7 = 3(\quad) + 23$

Check:

$4(\quad) + 12 = (\quad) + 39$

Check:

$13(\quad) + 28 = 5(\quad) + 52$

Solve each equation. Check each answer.

3. $11n - 6 = 9n$ $12y = 3y + 18$ $4x - 15 = x$

4. $4b + 17 = b + 23$ $2y + 14 = y + 17$ $5x - 7 = 2x + 33$

5. $9x + 32 = 4x + 82$ $8w - 13 = w + 50$ $3z + 14 = 2z + 14$

6. $\frac{3}{5}n - 2 = \frac{1}{5}n + 3$ $x + 8 = \frac{1}{4}x + 32$ $\frac{7}{8}y - 9 = \frac{3}{4}y + 5$

Applying Your Skills

....

The following examples will help you understand how an equation is written to represent information given in more complicated word problems.

EXAMPLE 1 Five times Blake's age minus 2 is equal to 2 times his age plus 10. How old is Blake?

    STEP 1 Let x equal Blake's age.

    STEP 2 Write an equation that represents the $5x - 2 = 2x + 10$
              information given in the problem.

    STEP 3 Solve the equation.

        a. Add 2 to each side. $5x - 2 + 2 = 2x + 10 + 2$

           Simplify. $5x = 2x + 12$

        b. Subtract $2x$ from each side. $5x - 2x = 2x - 2x + 12$

           Simplify. $3x = 12$

        c. Divide each side by 3. $\dfrac{3x}{3} = \dfrac{12}{3}$

           Simplify. $x = 4$

ANSWER: Blake is 4 years old.

EXAMPLE 2 Martin has money in a savings account. If he adds $200 each month for the next 8 months, he will have a total that is three times as much as he has now, not counting any interest. How much money is in Martin's account now?

    STEP 1 Let x = amount in Martin's account now.
              $x + \$1,600$ = amount in Martin's account in 8 months ($\$200 \times 8 = \$1,600$)
                    $3x$ = three times the amount he has now

    STEP 2 Write an equation for the problem. $x + \$1,600 = 3x$

    STEP 3 Solve the equation.

        a. Subtract x from each side. $x - x + \$1,600 = 3x - x$

           Simplify. $\$1,600 = 2x$

        b. Divide each side by 2. $\dfrac{\$1,600}{2} = \dfrac{2x}{2}$

           Simplify. $\$800 = x$
                                           or $x = \$800$

ANSWER: Martin has $800 in his account now.

(**Note:** Sometimes the variable ends up on the right side of the equals sign. Don't let this confuse you. Just solve the equation and, as your final step, write the variable on the left side.)

Solve the following problems using equations.

1. Four times a number plus 2 is equal to 2 times the same number plus 8.

2. Three fourths of an amount plus $150 is equal to twice the original amount.

3. Three numbers add up to 264. The second number is twice as large as the first. The third number is three times as large as the first. What are the three numbers?

4. Anna and Jo went to lunch. The total bill is $8.35. Anna's lunch cost $0.65 more than Jo's. How much is each person's share?

5. Sally is thinking of a number. She says, "If you divide the number by 4 and add 11 to the quotient, the number that results is equal to three times the original number." Which equation can you use to find Sally's number (n)?

    a. $\frac{n}{4} = 3n + 11$ c. $\frac{n+4}{11} = 3n$ e. $4n + 11 = 3n$

    b. $\frac{n+11}{4} = 3n$ d. $\frac{n}{4} + 11 = 3n$

6. Jean and Chris run a cleaning service. Jean provides all the supplies and earns $200 more each month than Chris. In July, the two earned a total of $2,400. Of this amount, how much should each person receive?

7. Consecutive integers are whole numbers that follow one another: for example, 7, 8, and 9. If the sum of 3 consecutive integers is 39, what are the integers?
    (**Hint:** Let $x =$ 1st integer; $x + 1 =$ 2nd integer; and $x + 2 =$ 3rd integer.)

8. Allison saves stamps. If she saves 25 stamps each month for the next year, she'll end up with 50% more stamps than she has now. How many stamps does Allison have now?

9. Lita makes twice as much profit on each blouse she sells as she does on each skirt. When she sells a blouse and skirt together, she makes a total profit of $10.50. How much profit does Lita make on each item?

10. For every dollar that Diva makes, her sister earns $1.25. Last month their total monthly income was $2,140. Which equation below can be used to determine Diva's income last month (x)?

    a. $x + \frac{4}{3}x = \$2,140$ c. $x + \frac{5}{4}x = \$2,140$ e. $x + x + \$0.75 = \$2,140$

    b. $x + x = \frac{\$2,140}{\$1.25}$ d. $x + x + \$1.25 = \$2,140$

Learning About Proportion

A **proportion** is made up of two equal ratios. For a review of ratios, reread pages 104–105. If you mix 2 cups of pineapple juice with 3 cups of orange juice, the ratio of pineapple juice to orange juice is $\frac{2}{3}$. You get the same-flavored mixture by adding 6 cups of pineapple juice to 9 cups of orange juice. The two mixtures have the same flavor because the ratios of ingredients are equal: $\frac{2}{3} = \frac{6}{9}$. The ratios form a proportion.

You write a proportion in symbols in either of two ways.

1. With colons, $2:3 = 6:9$

2. As equal fractions, $\frac{2}{3} = \frac{6}{9}$

You read a proportion as two equal ratios connected by the word *as*.

$2:3 = 6:9$ is read "2 is to 3 *as* 6 is to 9."

$\frac{2}{3} = \frac{6}{9}$ is read "2 is to 3 *as* 6 is to 9."

In a proportion, the **cross products** are equal. To find the cross products, cross multiply. Multiply each numerator by the opposite denominator.

Cross Multiplication

$$\frac{2}{3} \bowtie \frac{6}{9}$$

Equal Cross Products

$$2(9) = 3(6)$$
$$18 = 18$$

Many proportion problems ask you to find a missing number in a proportion. To solve this type of problem, use a variable to represent the missing number (x or another letter), and cross multiply. Then solve the equation.

EXAMPLE 1 Find the missing term: $\square:4 = 12:16$

STEP 1 Write the proportion as equal fractions, with the unknown value written as x.
$$\frac{x}{4} = \frac{12}{16}$$

STEP 2 Cross multiply.
$$16x = 4(12)$$
$$16x = 48$$

STEP 3 Solve for x.
$$x = \frac{48}{16} = 3$$

ANSWER: $x = 3$

EXAMPLE 2 Find the missing term: $\frac{7}{3} = \frac{28}{x}$

STEP 1 Cross multiply.
$$7x = 3(28)$$
$$7x = 84$$

STEP 2 Solve for x.
$$x = \frac{84}{7} = 12$$

ANSWER: $x = 12$

Cross multiply to see if each pair of fractions forms a proportion. (Remember: In a proportion, the cross products are equal.) Circle Yes if it is a true proportion. Circle No if no proportion is formed.

1. $\frac{3}{5} \overset{?}{=} \frac{20}{35}$ Yes   No $\frac{8}{5} \overset{?}{=} \frac{24}{16}$ Yes   No $\frac{11}{16} \overset{?}{=} \frac{33}{48}$ Yes   No

2. $\frac{4}{3} \overset{?}{=} \frac{28}{21}$ Yes   No $\frac{5}{6} \overset{?}{=} \frac{32}{36}$ Yes   No $\frac{8}{7} \overset{?}{=} \frac{56}{49}$ Yes   No

Write the following proportions as two equal fractions.

3. **a.** Five is to six as fifteen is to eighteen.   **b.** Three is to two as nine is to six.

4. $5:15 = 1:3$ $4:3 = 32:24$ $x:9 = 24:29$ $4:5 = 28:y$

Find the missing term in each proportion. When cross multiplying, write the product containing the variable to the left of the = sign.

5. $\frac{x}{4} = \frac{6}{8}$ $\frac{6}{15} = \frac{12}{y}$ $\frac{3}{n} = \frac{12}{8}$ $\frac{20}{25} = \frac{x}{5}$

6. $\frac{h}{9} = \frac{18}{27}$ $\frac{6}{x} = \frac{18}{12}$ $\frac{10}{6} = \frac{15}{y}$ $\frac{15}{16} = \frac{x}{64}$

7. $9:12 = 15:x$ $14:10 = y:5$ $10:6 = 5:h$ $3:8 = x:32$

8. $\square:28 = 3:7$ $4:\square = 12:15$ $3:2 = \square:48$ $7:16 = 21:\square$

Problem Solver: Using Proportions to Solve Word Problems

You can use proportions to solve word problems involving comparisons. As in ratio problems, be sure to write the terms of a proportion in the order stated in each problem.

EXAMPLE 1 A light blue paint is made by mixing 2 parts of blue paint to 5 parts of white paint. How many gallons of blue paint should be mixed with 24 gallons of white paint to make this light blue color?

STEP 1 Write a proportion where each ratio is $\frac{\text{amount of blue}}{\text{amount of white}}$.

Let x stand for the unknown number of gallons of blue.

Required ratios → $\frac{2}{5} = \frac{x}{24}$ ← Ratio of mixture being made

STEP 2 Cross multiply; then solve the resulting equation for x.

For simplicity, write $5x$ to the left of the = sign.

$$\frac{2}{5} = \frac{x}{24}$$

$$5x = 2(24) = 48$$

$$x = \frac{48}{5} = 9.6 \text{ gallons}$$

ANSWER: 9.6 gallons

Proportions may simplify your work with multistep rate problems.

EXAMPLE 2 On the first day of her trip, Pam drove 480 miles in 8 hours. Driving at the same rate, how far can Pam drive in $5\frac{1}{2}$ hours on the second day?

Instead of first finding the rate, you can use a proportion to solve this problem.

STEP 1 Write a proportion where each ratio is $\frac{\text{miles}}{\text{hours}}$.

Let x stand for the unknown mileage of the second day. (Notice that a number in the proportion itself may contain a fraction.)

First-day results → $\frac{480}{8} = \frac{x}{5\frac{1}{2}}$ ← Second-day results

STEP 2 Cross multiply; then solve the resulting equation for x.

$$\frac{480}{8} = \frac{x}{5\frac{1}{2}}$$

$$8x = 480(5\frac{1}{2}) = 2{,}640$$

$$x = \frac{2{,}640}{8} = 330 \text{ miles}$$

ANSWER: 330 miles

Solve the following problems by using proportions. (Notice that you can also solve several of these problems without using proportions.)

1. A tropical punch recipe calls for 3 parts club soda to 4 parts juice. How much club soda should be added to 3 quarts of juice to make the mixture?

2. Working for 8 hours, Brandon earned $159.20. At this pay rate, how much does Brandon earn in a 40-hour workweek?

3. If a 9-ounce piece of melon costs $1.98, what would be the price of a 13-ounce piece of the same type of melon?

4. Gloria drove her new car 255 miles on just 8 gallons of gas. Given similar driving conditions, how far can Gloria expect to drive on 12 gallons?

5. Which number correctly completes this statement: "If 9 defects are found in 270 television sets, you can expect to find _____ defects in 1,200 television sets."

   a. 20   b. 30   c. 40   d. 50   e. 60

6. While driving across Texas, Carol drove 392 miles on Monday in 7 hours. If she drives at the same rate, how far can Carol expect to drive on Tuesday in 4 hours and 30 minutes?

7. The Delta Company has 3 women managers for every 4 men managers. If Delta has 28 men managers, how many women managers does it have?

8. The length 25.4 centimeters equals exactly 10 inches. Knowing this, determine how many centimeters are in 25 inches.

9. Tough Guard epoxy is supposed to be mixed in the ratio of 3 parts hardener to 8 parts base. How many drops of hardener should you mix with 40 drops of base?

10. In a recent survey of 135 voters, 3 out of every 5 supported raising the gasoline tax. Which of the following proportions can be used to find the number of voters who are *not* in favor of raising this tax?

    a. $\frac{3}{5} = \frac{135}{x}$ b. $\frac{3}{7} = \frac{x}{135}$ c. $\frac{3}{5} = \frac{x}{135}$ d. $\frac{2}{5} = \frac{135}{x}$ e. $\frac{2}{5} = \frac{x}{135}$

Solving an Equation with Two Variables

So far in your study of algebra, you have learned about equations that contain a single variable. Now you'll learn how to solve and graph equations that contain two variables.

As an example, the equation $y = 3x + 2$ contains two variables, y and x.

dependent variable —↑   ↑— independent variable

Written as above, the variable to the left of the equals sign (y) is called the **dependent variable,** while the variable to the right (x) is called the **independent variable.**

An equation containing two variables has more than one solution. The value of the dependent variable *depends* on the value you choose for the independent variable. In $y = 3x + 2$, each y value depends on a chosen x value.

To solve an equation with two unknowns, follow these two steps.

> **STEP 1** Choose several values for the independent variable.
>
> **STEP 2** Substitute each chosen value of the independent variable into the original equation; then find the matching value of the dependent variable.

To keep the solutions in order, write them in a **Table of Values.**

EXAMPLE   Solve for y in the equation $y = 3x + 2$ for $x = 0, 1, 2,$ and 3.

> **STEP 1** Write the chosen x values in the Table of Values.
>
> **STEP 2** Substitute each x value into the equation $y = 3x + 2$
>
> For each x value, find the matching y value.
>
> **STEP 3** Fill in the Table of Values for y.

Choose x value.	Solve for y value.	Table of Values
x value	$y = 3x + 2$	

x value	Solve for y value	x	y
a. $x = 0$	$y = 3(0) + 2 = 2$	0	2
b. $x = 1$	$y = 3(1) + 2 = 5$	1	5
c. $x = 2$	$y = 3(2) + 2 = 8$	2	8
d. $x = 3$	$y = 3(3) + 2 = 11$	3	11

In each equation, identify both the dependent and the independent variable.

1. a. $y = 7x$

b. $c = 3s - 5$

c. $m = \frac{2}{3}n + 6$

dependent variable: _____

independent variable: _____

dependent variable: _____

independent variable: _____

dependent variable: _____

independent variable: _____

In each equation, find the value of the *dependent variable* for the given value of the *independent variable*.

2. $y = 4x$

If $x = 3$, then $y =$ _____.

$a = 5b + 9$

If $b = 0$, then $a =$ _____.

$m = 4n - 8$

If $n = 5$, then $m =$ _____.

Complete each Table of Values by solving each equation for the given values of the independent variable.

3. $y = x + 3$

Table of Values

x	y
0	
1	
2	
3	

4. $y = 2x - 2$

Table of Values

x	y
1	
2	
3	
4	

5. $p = 3r + 0.5$

Table of Values

r	p
1.5	
2	
2.5	
3	

6. $c = 4n + 1$

Table of Values

n	c
2	
4	
6	
8	

7. $r = \frac{1}{2}s + 6$

Table of Values

s	r
0	
3	
6	
9	

8. $m = \frac{2}{3}n - 1$

Table of Values

n	m
3	
6	
9	
12	

Becoming Familiar with a Coordinate Grid

The Coordinate Grid

To graph an equation, you need to become familiar with a **coordinate grid.** A coordinate grid is formed by combining a vertical number line, called the **y-axis,** with a horizontal number line, called the **x-axis.** (To review number lines, reread pages 22–23.)

The point at which the two axes meet is called the **origin** of the grid. The origin has a value of 0 for both axes.

Reading Points on a Coordinate Grid

Every point on a grid has **coordinates,** two numbers that tells its position.

- The *x-coordinate* tells how far the point is from the y-axis. Positive x indicates the point is to the right of the y-axis. Negative x indicates the point is to the left of the y-axis.
- The *y-coordinate* tells how far the point is from the x-axis. Positive y indicates the point is above the x-axis. Negative y indicates the point is below the x-axis.
- Coordinates are usually written in parentheses, the x-coordinate written first, followed by the y-coordinate: (x-coordinate, y-coordinate).

EXAMPLE   On the grid below, point A = (3, 4).

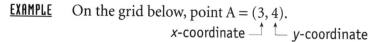

x-coordinate ⌐ ⌐ y-coordinate

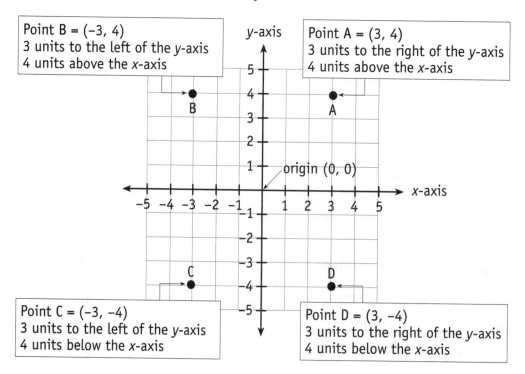

Point B = (−3, 4)
3 units to the left of the y-axis
4 units above the x-axis

Point A = (3, 4)
3 units to the right of the y-axis
4 units above the x-axis

Point C = (−3, −4)
3 units to the left of the y-axis
4 units below the x-axis

Point D = (3, −4)
3 units to the right of the y-axis
4 units below the x-axis

Plotting Points on a Coordinate Grid

To plot a point on a coordinate grid, follow these steps.

STEP 1 Locate the *x* value on the *x*-axis.

STEP 2 From this point, move directly up (for a positive *y* value) or directly down (for a negative *y* value) to the point that's directly across from the *y*-axis.

EXAMPLE To plot point A = (4, 3), find the coordinate point 4 on the *x*-axis. Then move directly up 3 spaces to the point that is across from 3 on the *y*-axis. Plot the point; then label it with an A or with its coordinates.

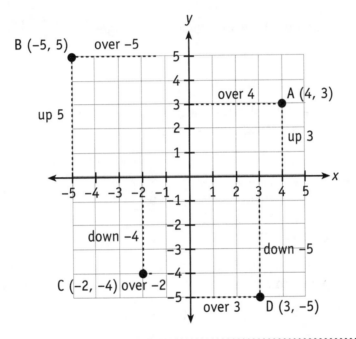

....

1. Identify the coordinates of each point.

2. Plot the points whose coordinates are given.

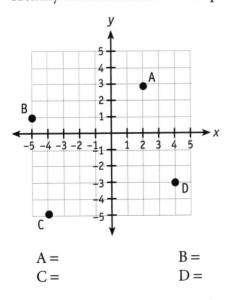

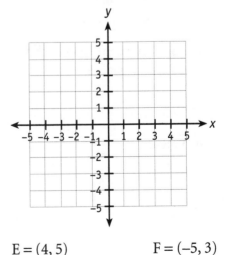

A =   B =

C =   D =

E = (4, 5)   F = (−5, 3)

G = (−2, −3)   H = (4, −5)

Graphing a Linear Equation

The equation $y = 2x + 1$ is an example of a **linear equation.** When solutions of a linear equation are plotted on a coordinate grid, they always lie on a straight line. Drawing this line of solutions is called **graphing the equation.**

To graph a linear equation, follow these steps.

STEP 1 Choose three values for the independent variable. Then determine the matching values for the dependent variable and write each pair of values in a Table of Values.

STEP 2 Plot the three points on a coordinate grid; then connect them with a line extending to the edges of the grid.

> **Math Tip**
> Only two points are needed to draw a line, but the third point serves as a check that you plotted the first two points correctly.

EXAMPLE Graph the equation $y = 2x + 1$.

STEP 1 Let $x = 0$, 1, and 2. Solve the equation $y = 2x + 1$ for these three values. Make a Table of Values.

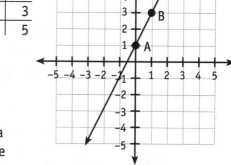

Table of Values

x value	$y = 2x + 1$
$x = 0$	$y = 2(0) + 1 = 1$
$x = 1$	$y = 2(1) + 1 = 3$
$x = 2$	$y = 2(2) + 1 = 5$

x	y
0	1
1	3
2	5

STEP 2 Write the pairs of values in the Table of Values as points to be plotted.

A = (0, 1)   B = (1, 3)   C = (2, 5)

STEP 3 Plot the points, then connect them with a straight line extending to the edges of the grid.

Graph each equation on the grid at the right. Use the listed values of each independent variable.

1. $y = 2x - 4$

Table of Values

x	y
2	
3	
4	

2. $y = \frac{1}{2}x + 1$

Table of Values

x	y
0	
2	
4	

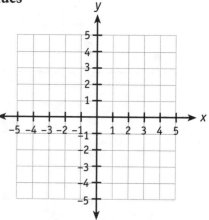

3. Blue Cab Company charges a $1.25 pickup charge plus $1.25 per mile for local service. If *f* stands for fare (total cost of the trip) and *m* for miles, the fare equation is written as follows.

$$f = \$1.25m + \$1.25$$

a. Using the fare equation, determine the fare of a cab trip of 8 miles.

b. Fill in the Table of Values for the *m* values listed, then graph the fare equation.

Table of Values

m	f
1	
3	
5	

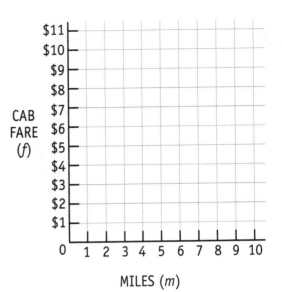

c. Using the graph, estimate the distance a fare of $10.00 will take you.

4. In the United States, we use both the **Fahrenheit**-scale thermometer and the **Celsius**-scale thermometer. The following equation gives the Fahrenheit temperature (°F) for a given Celsius temperature (°C).

$$°F = \frac{9}{5}°C + 32°$$

a. Water boils at 100°C. Using the temperature equation, determine the temperature in °F of the boiling point of water.

b. Fill in the Table of Values for the °C values listed; then graph the temperature equation.

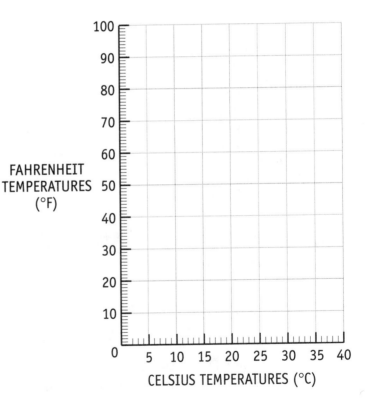

Table of Values

°C	°F
15	
25	
35	

c. Water freezes at 32°F. From the graph, estimate the equivalent Celsius temperature.

Finding Slope

You are already familiar with the concept of **slope.** When you walk uphill, you walk up a slope. When you walk downhill, you walk down a slope. A graphed line, like a hill, has a slope.

The slope of a line is given as a number. When you move between two points on the line, the slope is found by dividing the change in y values by the corresponding change in x values.

$$\text{Slope of a line} = \frac{\text{Change in } y \text{ values}}{\text{Change in } x \text{ values}}$$

- A line that goes up from left to right has a *positive slope.*
- A line that goes down from left to right has a *negative slope.*

In Graph A, the slope of the line is 2.

Graph A

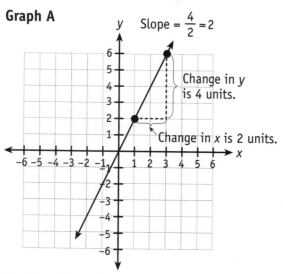

Slope $= \dfrac{4}{2} = 2$

Change in y is 4 units.

Change in x is 2 units.

The line above has *positive slope* because it goes *up* from left to right.

In Graph B, the slope of the line is −1.

Graph B

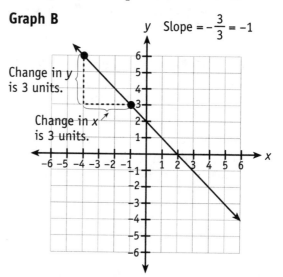

Slope $= -\dfrac{3}{3} = -1$

Change in y is 3 units.

Change in x is 3 units.

The line above has *negative slope* because it goes *down* from left to right.

Zero Slope and Undefined Slope

Two types of lines have neither positive nor negative slope.

- A horizontal line has *zero slope.* The x-axis (or any horizontal line) is a line with 0 slope.

- A vertical line has an *undefined slope.* The concept of slope does not apply to a vertical line. The y-axis (or any vertical line) is a line with undefined slope.

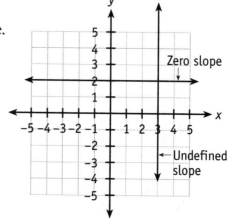

Zero slope

Undefined slope

1. Name the slope of each line as *positive, negative, zero,* or *undefined.*

   Line A:_____   Line C:_____

   Line B:_____   Line D:_____

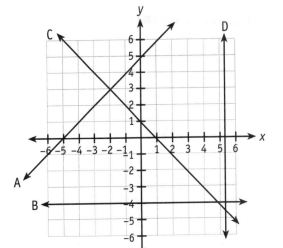

2. Find the numerical value of the slope of line E.

   Slope of Line E is _____.

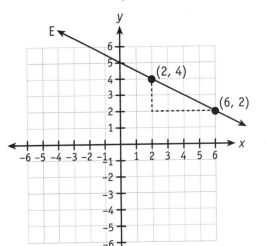

3. **a.** Find the numerical value of the slope of line M.
   b. What are the coordinates of the *x-intercept*, the point where line M crosses the *x*-axis?
   c. What are the coordinates of the *y-intercept*, the point where line M crosses the *y*-axis?

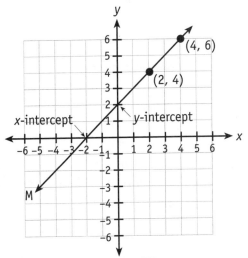

4. By subtracting coordinates, determine the slope of the line that passes through each of the following pair of points. Part a is completed as an example.

   a. $(2, 4)$ and $(5, 7)$

   $$\text{slope} = \frac{\text{change in } y}{\text{change in } x} = \frac{7 - 4}{5 - 2} = \frac{3}{3} = 1$$

   b. $(0, 0)$ and $(3, 4)$

   c. $(2, 5)$ and $(6, 8)$

Algebra Review

Solve each problem below.

1. What rule do you follow to solve the following equation for *x*?

    $4x = 36$

   a. Subtract 4 from each side.
   b. Multiply each side by 4.
   c. Subtract 4 from each side.

   d. Divide each side by 4.
   e. Divide each side by 36.

2. During a sale, Jan paid 75% of the original price of a sweater. If Jan paid $27 for the sweater, which equation below can be solved to find the original price (*p*)?

   a. $0.75 = \$27p$
   b. $0.75p = \$27$
   c. $0.25p = \$27$

   d. $0.25 = p - \$27$
   e. $0.75 = p - \$27$

3. What is the solution of the following equation?

    $3x + 9 = 27$

4. Which equation below represents the information given in the following statement?

    "Four times *y* minus 2 is equal to 2 times *y* plus 6."

   a. $4y - 2 = 2y + 6$
   b. $4y + 2 = 2y + 6$
   c. $4y - 2 = 2y - 6$

   d. $4y - 2 = 2y - 6$
   e. $4y + 2y = 6 - 2$

5. If you subtract 8 from 7 times a certain number, you get 34. What is the number?

6. Karen and Robin share monthly expenses. However, Robin pays $50 less each month than Karen. How much is Karen's share of January's $780 total?

7. Frank, Albert, and Carmine worked together on a job. Frank earned $40 less than Albert. Carmine earned $50 more than Albert. If the three men together earned $400 for the job, how much did Frank earn?

   a. $55   **b.** $70   **c.** $90   **d.** $105   **e.** $130

8. **Consecutive even numbers** are even whole numbers that follow one another: for example, 4, 6, and 8. If the sum of three consecutive even numbers is 42, which equation can be used to find the smallest (x) of the three numbers?

   a. $x + 2x + 4x = 42$

   b. $x + 2x = 4x + 2 = 42$

   c. $x + x + x = 42$

   d. $x + x + 2 + x + 4 = 42$

   e. $x + 2 + 4 = 42$

9. What is the missing term (value of y) in the following proportion?

   $$\frac{12}{9} = \frac{48}{y}$$

10. Elvira walked 3 miles in 40 minutes. If she maintains this rate, about how many minutes will it take Elvira to walk 7 miles?

11. In a local poll of 98 shoppers, 5 out of every 7 said they prefer low-fat milk over whole milk. Which proportion can be used to find the number of shoppers who said they prefer low-fat?

    a. $\frac{7}{12} = \frac{x}{98}$

    b. $\frac{5}{7} = \frac{98}{x}$

    c. $\frac{5}{12} = \frac{x}{98}$

    d. $\frac{7}{12} = \frac{98}{x}$

    e. $\frac{5}{7} = \frac{x}{98}$

12. In the equation $y = 4x - 7$, what is the value of y when $x = 2$?

13. Which of the lettered points on the grid has the coordinates $(-3, 2)$?

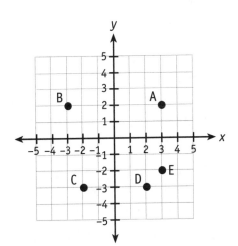

14. The line on the grid contains the point $(4, 2)$. Which of the following is the equation of this line?

    a. $y = 2x + 1$

    b. $y = \frac{1}{3}x + 2$

    c. $y = \frac{1}{2}x - 2$

    d. $y = 3x - 1$

    e. $y = \frac{1}{4}x + 1$

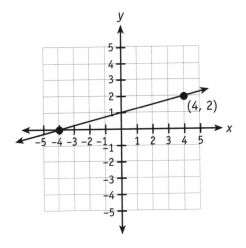

GEOMETRY

Geometry includes the study of angles and triangles and the study of perimeter, area, and volume.

Introducing Angles

A **ray** is a line with an **endpoint** and extends in one direction.
An **angle** is formed when two rays meet at the same endpoint.
- The rays that form the angle are called the **sides** of the angle.
- The point where the rays meet is called the **vertex** of the angle.

The symbol for angle is ∠.

An arc ⌒ is usually drawn to show which of two possible angles is meant.

For example,  might represent either  or .

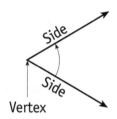

endpoint

Side

Side

Vertex

Labeling Angles

An angle is labeled (named) in one of the three ways shown below.

By Three Letters

A

B

C

∠ABC *or* ∠CBA

When you are using three letters, the vertex letter is always written as the middle letter.

By a Single Vertex Letter

B

∠B

A single vertex letter or number can be placed either outside or inside the angle.

By a Number

2

∠2

Write a letter or number name for each angle pictured below.

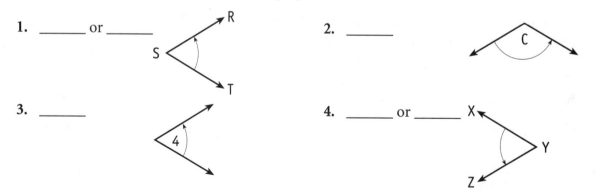

1. _____ or _____

R

S

T

2. _____

C

3. _____

4

4. _____ or _____

X

Y

Z

Measuring Angles

The size of an angle depends only on the opening between its sides. This opening is measured in units called **degrees.** The symbol for degrees is °. A 45-degree angle is written as 45°.

Degrees can be thought of as parts of a divided circle. A whole circle contains 360°.

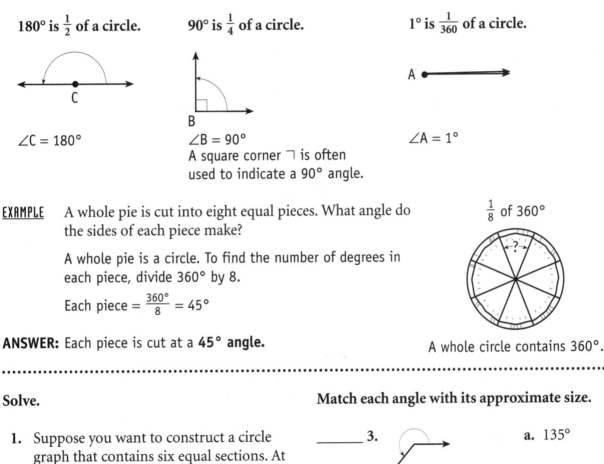

180° is $\frac{1}{2}$ of a circle.

∠C = 180°

90° is $\frac{1}{4}$ of a circle.

∠B = 90°
A square corner ⌐ is often used to indicate a 90° angle.

1° is $\frac{1}{360}$ of a circle.

∠A = 1°

EXAMPLE A whole pie is cut into eight equal pieces. What angle do the sides of each piece make?

A whole pie is a circle. To find the number of degrees in each piece, divide 360° by 8.

Each piece = $\frac{360°}{8}$ = 45°

$\frac{1}{8}$ of 360°

A whole circle contains 360°.

ANSWER: Each piece is cut at a **45° angle.**

Solve.

1. Suppose you want to construct a circle graph that contains six equal sections. At what angle should you draw the sides of each section?

2. A spinner for a new board game is equally likely to stop in any one of 10 same-size sections. At what angle do the sides of each section meet?

Match each angle with its approximate size.

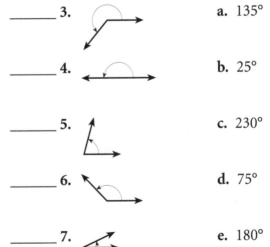

_____ 3.

_____ 4.

_____ 5.

_____ 6.

_____ 7.

a. 135°

b. 25°

c. 230°

d. 75°

e. 180°

Angle Relationships

Types of Angles

An angle can be classified by its size.

Acute Angle

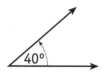

More than 0° but less than 90°

Right Angle

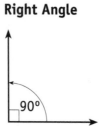

Exactly 90°

Obtuse Angle

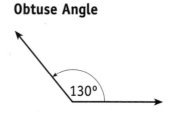

More than 90° but less than 180°

Straight Angle

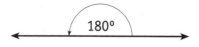

Exactly 180°

Reflex Angle

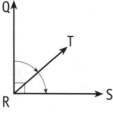

More than 180° but less than 360°

Pairs of Angles

Angles can be added to form a single larger angle.

Two angles that add to 90° are
called **complementary angles.**

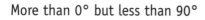

∠QRT + ∠TRS are complementary.
∠QRT is the *complement* of ∠TRS.

Two angles that add to 180° are
called **supplementary angles.**

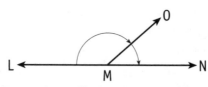

∠LMO and ∠OMN are supplementary.
∠LMO is the *supplement* of ∠OMN.

EXAMPLE   As shown at the right, ∠ABC is a right angle (90°).
What is ∠ABD if ∠DBC is 47°?

∠ABD = 90° − 47° = 43°

ANSWER: ∠ABD = 43°

∠ABD and ∠DBC are
complementary angles.

∠ABD + ∠DBC = 90°

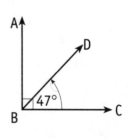

Name each angle below as acute, right, obtuse, straight, or reflex.

1. _____

29°

2. _____

300°

3. _____

140°

4. _____

180°

5. _____

90°

6. _____

68°

Find the value of each angle indicated below.

7. ∠DEG = _____

D
G
37°
E   F

8. ∠BCE = _____

E
57°
B   C   D

9. ∠PNO = _____

M
O   52°
P   N

Solve each problem below.

10. The diagonal brace in a fence gate makes a 38° angle with the bottom crosspiece. What angle does the brace make with the vertical side piece?

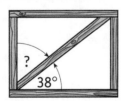

11. The side of an A-frame house makes a 125° angle with the ground. How large is the acute angle that this side makes with the floor of the house?

12. A ladder is leaning against the side of a house. The ladder makes an acute angle of 64° with the ground. What is the value of the obtuse angle the ladder makes with the ground?

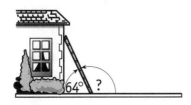

13. The two sides of a picture frame are cut so that they can be joined at the ends. At what angle must each side be cut so that the angle made by the joined sides is 90°?

Angles Formed by Intersecting Lines

Many angles are formed by lines crossing one another. Here are two examples that are often used as the basis of test questions.

Vertical Angles

Vertical angles are the angles opposite each other when two straight lines cross.

> Vertical angles are equal.

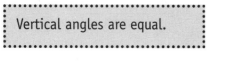

$\angle 1 = \angle 3$ and $\angle 2 = \angle 4$

When two lines meet (or cross) at a right angle, the lines are called **perpendicular lines.** Vertical angles formed by crossing perpendicular lines are all 90° angles. The symbol for perpendicular is ⊥. At the right, line A is perpendicular to line B, or A ⊥ B.

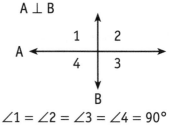

$\angle 1 = \angle 2 = \angle 3 = \angle 4 = 90°$

Parallel Lines Cut by a Transversal

Parallel lines are straight lines that run side by side and never cross. The symbol for parallel is ∥. Below, line C is parallel to line D, or C ∥ D.

A **transversal** is a third line that crosses two other lines.

> When two parallel lines are cut by a transversal
>
> • the four acute angles are equal
>
> • the four obtuse angles are equal
>
> • each acute angle is supplementary to each obtuse angle

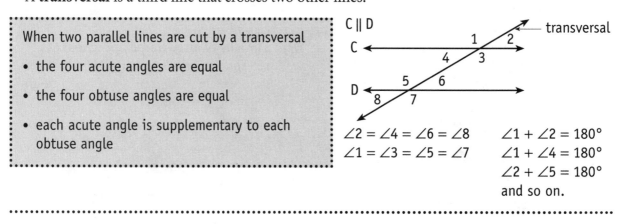

$\angle 2 = \angle 4 = \angle 6 = \angle 8$ $\angle 1 + \angle 2 = 180°$
$\angle 1 = \angle 3 = \angle 5 = \angle 7$ $\angle 1 + \angle 4 = 180°$
$\angle 2 + \angle 5 = 180°$
and so on.

Find the value of each angle as indicated.

1. $\angle 9 =$ _____

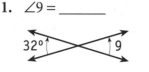

2. $\angle 10 =$ _____

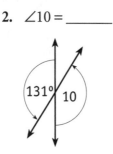

3. $\angle 11 =$ _____

L ⊥ M

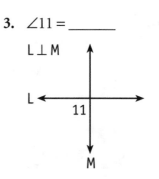

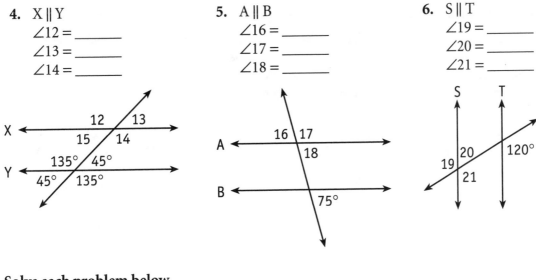

4. X ∥ Y

∠12 = _____

∠13 = _____

∠14 = _____

5. A ∥ B

∠16 = _____

∠17 = _____

∠18 = _____

6. S ∥ T

∠19 = _____

∠20 = _____

∠21 = _____

Solve each problem below.

7. Smith Road intersects Highway 34 as shown below. What are the values of the three unmeasured angles?

∠1 = _____

∠2 = _____

∠3 = _____

8. Roosevelt Avenue crosses 5th and 6th Streets as shown below. If 5th Street is parallel to 6th Street, what is the value of the acute angle that Roosevelt makes with 6th Street?

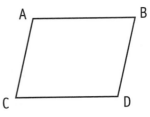

9. The letter E is formed by joining parallel lines with a transversal. Name four other capital letters that are formed with parallel lines and a transversal.

E

10. A **parallelogram** is a four-sided figure that has two pairs of parallel sides. Using your knowledge of angles, find the sum of the four angles of a parallelogram.

(**Hint:** Each obtuse angle is supplementary to each acute angle.)

∠A + ∠B + ∠C + ∠D = _____

Working with Triangles

A **triangle** is a closed **plane** (flat) **figure** that has three sides and three angles. Triangles are named with three letters, one placed at the vertex of each angle. Each angle can be named by a single letter. Each side can be named by a pair of letters. The symbol for triangle is Δ.

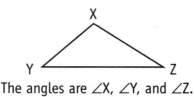

The angles are ∠X, ∠Y, and ∠Z.
The sides are XY, XZ, and YZ.
ΔXYZ

Types of Triangles

Below are the four most common types of triangles.

Equilateral Triangle

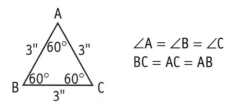

∠A = ∠B = ∠C
BC = AC = AB

An **equilateral triangle** has three equal sides and three equal angles. Each angle is 60°.

Isosceles Triangle

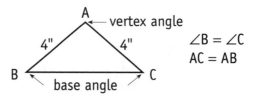

∠B = ∠C
AC = AB

An **isosceles triangle** has at least two equal sides and two equal angles called **base angles.** The third angle is called the **vertex angle.**

Scalene Triangle

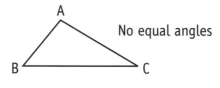

No equal angles

A **scalene triangle** has no equal sides and no equal angles.

Right Triangle

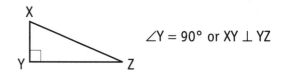

∠Y = 90° or XY ⊥ YZ

An isosceles or scalene triangle that has a 90° angle is also called a **right** triangle.

Angle Relationships in a Triangle

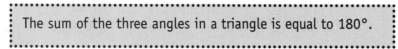

The sum of the three angles in a triangle is equal to 180°.

EXAMPLE   What is the value of ∠B in ∠ABC?

STEP 1   Add ∠A and ∠C.
∠A + ∠C = 67°

STEP 2   To find ∠B, subtract 67° from 180°.
∠B = 180° − 67° = 113°

ANSWER: ∠B = 113°

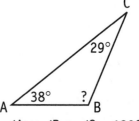

∠A + ∠B + ∠C = 180°

Name each triangle below as equilateral, isosceles, or scalene. Circle any right triangle.

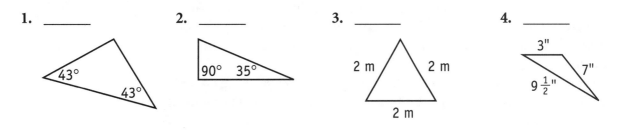

1. _____

2. _____

3. _____

4. _____

Find the value of each angle indicated below.

5. ∠B = _____

6. ∠G = _____

7. ∠Y = _____

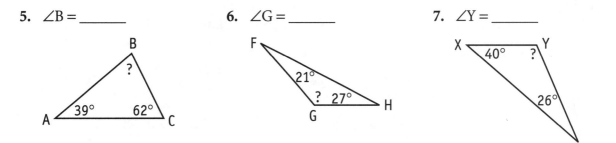

8. What is the roof angle (∠x) in the drawing below?

9. Referring to the drawing below, how large is the acute angle that the support makes with the shelf?

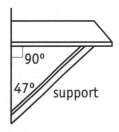

10. Look carefully at the drawing below and answer the following questions.

    a. At what acute angle does the Amtrak line cross Maple Street (∠3)?

    b. How large is ∠5, one of the acute angles that Smith's property line makes with the Amtrak line? (**Hint:** As a first step, determine the value of ∠3.)

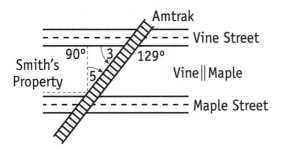

11. ∠A and ∠B are supplementary angles. ∠A is four times as large as ∠B. Determine the size of each angle. (**Hint:** Let ∠A = 4x and ∠B = x. Let the sum = 180°; then solve.)

12. In ΔCDE, ∠C is the smallest angle. ∠D is twice as large as ∠C, and ∠E is three times as large as ∠C. Determine the size of ∠C, ∠D, and ∠E. (**Hint:** Let ∠C = x, ∠D = 2x, and ∠E = 3x. Let the sum = 180°; then solve.)

Problem Solver: Working with Similar Triangles

Similar triangles are triangles where each angle in one triangle equals an angle in the other triangle. Similar triangles have the same shape and differ only in the lengths of their sides. The symbol ~ stands for *is similar to*.

In similar triangles, the equal angles are called **corresponding angles** and the sides that are opposite the equal angles are called **corresponding sides.**

The lengths of corresponding sides can be written as a proportion.

<u>EXAMPLE 1</u> ΔABC and ΔDEF are similar triangles because they have three pairs of equal angles. You can write a proportion as follows.

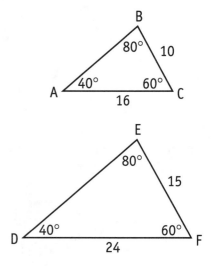

STEP 1 Identify the pairs of corresponding sides.

    a. AC and DF are corresponding sides. (Each is opposite an 80° angle.)

    b. BC and EF are corresponding sides. (Each is opposite a 40° angle.)

    c. AB and DE are corresponding sides. (Each is opposite a 60° angle.)

STEP 2 Write a proportion using two of the pairs.

ANSWER: $\frac{BC}{EF} = \frac{AC}{DF}$ or $\frac{10}{15} = \frac{16}{24}$

(**Note:** The numerator of each fraction is from one triangle, while each denominator is from the other triangle.)

Example 2 shows how to find an unknown length when working with similar triangles.

<u>EXAMPLE 2</u> ΔXYZ and ΔRST are similar triangles. The length of side YZ is unknown and is represented as *d*. Find this length.

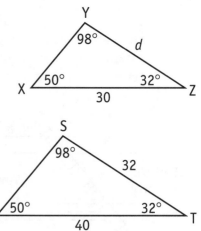

STEP 1 Write a proportion using the corresponding sides. To simplify your work, write the proportion so that *d* (the unknown length) is the numerator of the left-side fraction.

$$\frac{YZ}{ST} = \frac{XZ}{RT} \text{ or } \frac{d}{32} = \frac{30}{40}$$

STEP 2 Solve the proportion for *d*.

$$40d = 32(30) \text{ or } 960$$
$$d = 960 \div 40$$
$$d = 24$$

ANSWER: The length of YZ is 24.

Write and solve a proportion to find each unknown length.

1. △ABC ~ △DEF. What is length *d*?

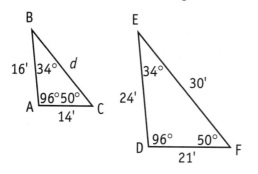

2. △HIJ ~ △RST. What is length *x*?

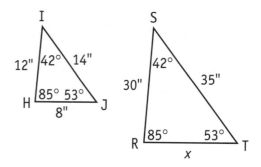

Use a proportion to solve each problem below.

3. Two trees stand side by side. The smaller tree is 12 feet high and casts a shadow of 20 feet. At the same time of day, the larger tree casts a shadow of 55 feet. How tall is the larger tree? (**Hint:** Similar triangles are formed by the trees, the ground, and the dotted lines that represent the sun's rays.)

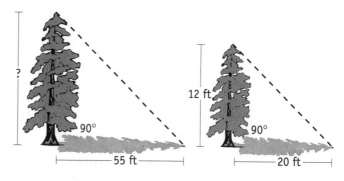

4. At her campsite, Aida estimated the distance (*d*) across Odell River. To do this, she walked off the distances as shown in her drawing at the right. Use her drawing to estimate the distance across the river.
(**Hint:** The two triangles are similar because the three pairs of corresponding angles are equal. Here's why:

- Each has a right angle (90°). These angles are indicated by ⌐.

- The vertical angles, labeled 1 and 2, are equal.

- The third angles, labeled 3 and 4, are also equal. Reason: If two triangles have two equal angles, the third angles are equal because the sum of the angles in each triangle must be 180°.)

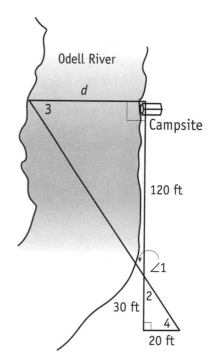

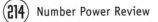

Understanding Squares and Square Roots

Squares

The **square** of a number is that number multiplied by itself. The square of 7 is $7 \times 7 = 49$. In symbols, you write the square of a number as a **base** and an **exponent.**

7×7 is written $7^2 \leftarrow$ exponent
$\quad\quad\quad\quad\quad\;\;\uparrow$ base

The exponent (2) tells how many times to write the base (7) as a **factor** (a number being multiplied). The exponent of a square is always 2. Read 7^2 as "7 squared" or "7 to the 2nd power." The value of 7^2 is 49. Here are three more examples.

Squared Number	As a base and an exponent	Read in words	Value
3×3	3^2	"three squared"	9
10×10	10^2	"ten squared"	100
$b \times b$	b^2	"b squared"	*

*A value for "variable squared" can be found only when you know the value of the variable. For example, if $b = 4$, then $b^2 = 16$; if $b = 9$, then $b^2 = 81$.

Square Roots

To find the **square root** of a number, you ask, "What number, when multiplied by itself, equals this number?"

What number times itself equals 49? The answer is 7, because $7 \times 7 = 49$. The symbol for square root is $\sqrt{}$. You write $\sqrt{49} = 7$.

The table at the right contains the squares of number from 1 to 15. These numbers 1, 4, 9, 16, and so on, are called **perfect squares** because their square roots are whole numbers.

The values in this table are used to find the square roots of these perfect squares. For example, if $c^2 = 121$, what is c? To determine c, find 121 on the table.

The table shows $11^2 = 121$. Thus, $\sqrt{121} = 11$, and $c = 11$.

Table of Perfect Squares		
$1^2 = 1$	$6^2 = 36$	$11^1 = 121$
$2^2 = 4$	$7^2 = 49$	$12^2 = 144$
$3^2 = 9$	$8^2 = 64$	$13^2 = 169$
$4^2 = 16$	$9^2 = 81$	$14^2 = 196$
$5^2 = 25$	$10^2 = 100$	$15^2 = 225$

Find the value of each squared number or expression below.

1. $5^2 =$

2. $8^2 =$

3. $10^2 =$

4. $(\frac{1}{2})^2 =$

5. $(3.5)^2 =$

6. $(\frac{2}{3})^2 =$

7. If $a = 6$, then $a^2 =$

8. If $x = 3.2$, then x^2

9. If $y = \frac{7}{8}$, then $y^2 =$

Use the Table of Perfect Squares to find each square root below.

10. $\sqrt{16} =$

11. $\sqrt{100} =$

12. $\sqrt{36} =$

13. $\sqrt{169} =$

14. $\sqrt{225} =$

15. $\sqrt{196} =$

16. If $a^2 = 25$, then $a =$

17. If $b^2 = 49$, then $b =$

18. If $x^2 = 121$, then $x =$

Find each value below. Find the value of each squared number before finding a sum or difference.

19. If $a = 5$ and $b = 6$, then $a^2 + b^2 =$

20. If $x = 7$ and $y = 4$, then $x^2 - y^2 =$

21. If $a = 3$ and $b = 4$, then $\sqrt{a^2 + b^2} =$

22. If $m = 10$ and $n = 8$, then $\sqrt{m^2 - n^2} =$

Estimate the value of each square root below. Two are done as examples. Then use a calculator to find a more precise value.*

23. $\sqrt{50} \approx 7$
(since $50 \approx 49$)
Calculator answer: ≈ 7.07

24. $\sqrt{35} \approx$

25. $\sqrt{90} \approx 9.5$
(since 90 is about halfway between 81 and 100)
Calculator answer: ≈ 9.49

26. $\sqrt{12} \approx$

27. $\sqrt{98} \approx$

28. $\sqrt{155} \approx$

*To find a square root with a calculator, enter the number and then press the $\boxed{\sqrt{}}$ key or $\boxed{\sqrt{x}}$ key.

The Pythagorean Theorem

In a *right triangle*, the side opposite the right angle (90°) is called the **hypotenuse.** The hypotenuse is the longest side.

The Greek mathematician Pythagoras discovered an important relationship between the hypotenuse and the two shorter sides. This relationship is called the **Pythagorean theorem.**

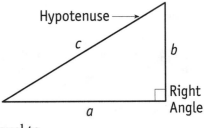

In words: In a right triangle, the square of the hypotenuse is equal to the sum of the squares of the other two sides.

In symbols: $c^2 = a^2 + b^2$ (using the labels on the triangle above)

If you know the lengths of the two shorter sides, you can use the Pythagorean theorem to find the length of the hypotenuse.

EXAMPLE   What is the length of the hypotenuse of ΔLMN?

STEP 1   Find the square of each shorter side.

  a. Substitute 3 for a and find a^2.
    $a^2 = 3^2 = 9$

  b. Substitute 4 for b and find b^2.
    $b^2 = 4^2 = 16$

STEP 2   Add the squares found in Step 1.

  $9 + 16 = 25$

STEP 3   To find c, take the square root of 25.

  $c = \sqrt{25} = 5$

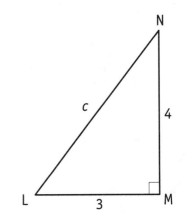

ANSWER: The length of the hypotenuse is 5.

You can also find the length of a side of a right triangle if you know the lengths of the hypotenuse and the other side. To do this, rewrite the Pythagorean theorem as follows.

In symbols: $a^2 = c^2 - b^2$

In words: Side a squared equals hypotenuse squared minus side b squared.

In the example above, if you know that the hypotenuse is 5 and that one side is 4, you can find the other side by squaring and subtracting.

$a^2 = 5^2 - 4^2$
$\quad = 25 - 16 = 9$
So, $a = \sqrt{9} = 3$ (as we already know).

Use the Pythagorean theorem to find the length of each unmeasured side below. (For problems 1–5, refer to the Table of Perfect Squares on page 214.)

1. $c =$ _____

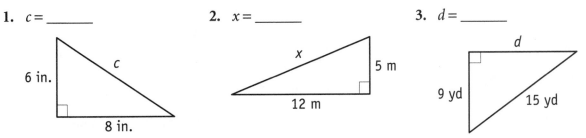

6 in.

c

8 in.

2. $x =$ _____

x

5 m

12 m

3. $d =$ _____

d

9 yd

15 yd

4. A **rectangle** is a four-sided figure with four right angles. A **diagonal** divides a rectangle into two right triangles. What is the length of the diagonal in the rectangle at right?

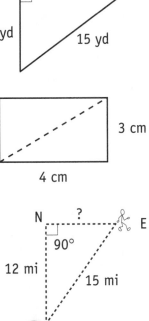

3 cm

4 cm

5. Jenny hiked 12 miles north and several miles east during the first day of her camping trip. If she is now 15 miles from her car, how many miles east did she hike?

N   ?   E

90°

12 mi

15 mi

Start

 Use your calculator for problems 6 and 7.

6. A ladder rests against the side of Kate's house. The bottom of the ladder is 8 feet from the house, and the top just reaches a balcony that's 15 feet above ground. How long is the ladder?

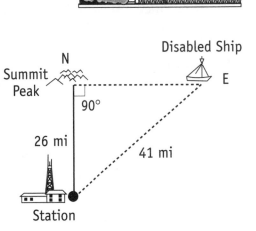

15 ft

?

8 ft

7. A disabled ship is 41 miles from the Coast Guard station. The ship is directly east of Summit Peak, which is 26 miles directly north of the station. To the nearest mile, how far is the ship from Summit Peak?

Disabled Ship

N

Summit Peak

E

90°

26 mi

41 mi

Station

Finding the Perimeter

The distance around a plane (flat) object is known as its **perimeter.** You find the perimeter of a many-sided object by adding the lengths of its sides. The symbol for perimeter is *P*. For squares, rectangles, and triangles, you often see the perimeter formulas written below.

Perimeters of Squares, Rectangles, and Triangles

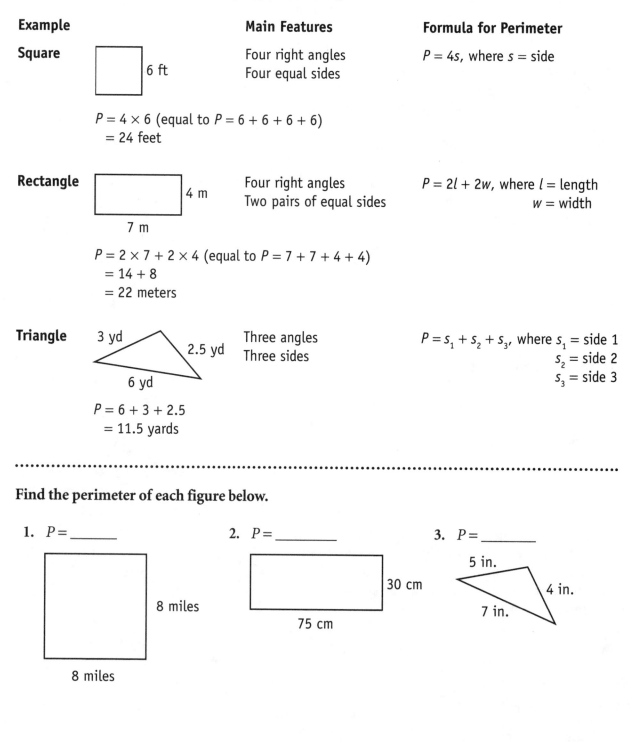

Example	Main Features	Formula for Perimeter
Square 6 ft	Four right angles Four equal sides	$P = 4s$, where s = side

$P = 4 \times 6$ (equal to $P = 6 + 6 + 6 + 6$)
$= 24$ feet

| **Rectangle** 4 m, 7 m | Four right angles
Two pairs of equal sides | $P = 2l + 2w$, where l = length
w = width |

$P = 2 \times 7 + 2 \times 4$ (equal to $P = 7 + 7 + 4 + 4$)
$= 14 + 8$
$= 22$ meters

| **Triangle** 3 yd, 2.5 yd, 6 yd | Three angles
Three sides | $P = s_1 + s_2 + s_3$, where s_1 = side 1
s_2 = side 2
s_3 = side 3 |

$P = 6 + 3 + 2.5$
$= 11.5$ yards

Find the perimeter of each figure below.

1. $P =$ _____

8 miles
8 miles

2. $P =$ _____

30 cm
75 cm

3. $P =$ _____

5 in.
4 in.
7 in.

Finding the Circumference of a Circle

A **circle** is a figure in which all the points are at an equal distance from the center. This distance is called the **radius.** The symbol for radius is *r*.

The distance across a circle is called the **diameter.** As you may know, the diameter passes through the center of the circle and is equal to twice the radius. The symbol for diameter is *d*.

The distance around a circle is called the **circumference.** The symbol for circumference is *C*.

The circumference is given by the formula $C = \pi d$, where π has the approximate value 3.14 or $\frac{22}{7}$ (π is a letter from the Greek alphabet. It is pronounced "pie").

$d = 2r$ or $r = \frac{1}{2}d$

> To compute the circumference of a circle, multiply π times the diameter.

EXAMPLE 1 What is the circumference of a circle that has a diameter of 7 feet?

To find the circumference, multiply $\frac{22}{7}$ by 7.

$C = \frac{22}{7} \times 7 = \frac{22}{\overset{}{\underset{1}{7}}} \times \frac{\overset{1}{7}}{1} = 22$

ANSWER: 22 feet

(**Note:** Use $\pi \approx \frac{22}{7}$ when the radius or diameter is a multiple of 7: 7, 14, 21, 28, and so on.)

EXAMPLE 2 Find the circumference of a circle that has a diameter of 2.5 yards.

Multiply 3.14 by 2.5.

$C = 3.14 \times 2.5 = 7.85$

ANSWER: 7.85 yards

(**Note:** Use $\pi \approx 3.14$ when the radius or diameter is given as a decimal.)

Use the formula $C = \pi d$ to solve each problem below.

1. The Community Wading Pool is in the shape of a circle. If the radius of the pool is 28 feet, what is the distance around the pool?

2. The diameter of a circular garden is 6.5 meters. To the nearest tenth of a meter, what is the circumference of this garden?

3. The tire on Frank's bicycle has a radius of $12\frac{1}{4}$ inches.

    a. What is the diameter of this tire?

    b. To the nearest inch, what is the tire's circumference?

4. A sandbox in the shape of a circle has a radius of 3.3 yards.

    a. What is the diameter of the sandbox?

    b. To the nearest yard, what is the circumference of this sandbox?

Finding the Area of a Square or a Rectangle

Area is a measure of a surface or a flat region. For example, to measure the size of a floor, you determine its area. The symbol for area is *A*.

Common area units are the **square inch** (sq in.), **square foot** (sq ft), and **square yard** (sq yd). In the metric system, the most common area units are the **square centimeter** (cm^2) and the **square meter** (m^2).

Sample Area Unit
1 square foot

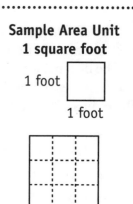

The area of a square is given by the formula $A = s^2$, where *s* stands for *side*.

> To compute the area of a square, multiply the side by itself.

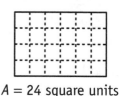

$A = 9$ square units

The area of a rectangle is given by the formula $A = lw$, where *l* stands for *length* and *w* stands for *width*.

> To compute the area of a rectangle, multiply its length by its width.

$A = 24$ square units

EXAMPLE 1 What is the area of a square that measures 3 feet on each side?

To compute this area, multiply 3×3.

$A = 3 \times 3 = 9$

3 ft ☐ 3 ft

ANSWER: 9 square feet

EXAMPLE 2 What is the area of a rectangle that is 11 feet long and 4 feet wide?

To find this area, multiply 11×4.

$A = 11 \times 4 = 44$

4 ft ▭ 11 ft

ANSWER: 44 square feet

Find the area of each figure.

1. *A* = _____

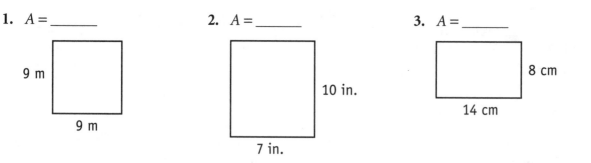

9 m
9 m

2. *A* = _____

10 in.
7 in.

3. *A* = _____

8 cm
14 cm

Solve each problem below. You may find it useful to draw a picture.

4. Linda is buying a piece of glass to fit inside a square picture frame. How many square inches of glass does she need if the space in which the glass fits measures 7.5 inches by 7.5 inches?

5. Brenda's bedroom floor measures $4\frac{1}{2}$ yards long and $3\frac{3}{4}$ yards wide. How many square yards of carpet will she need to buy to cover this floor?

6. The area of the top of a square table is 16 square feet. How long is each side of the table? (**Hint:** For a square, $s = \sqrt{A}$.)

7. The label on a gallon of interior wall paint reads "Coverage: 400 square feet." What length of 8-foot-high wall can be painted with one gallon of this paint? (**Hint:** For a rectangle, $l = A \div w$.)

8. Using square tiles that measure 9 inches on each side, Ben is tiling his shop floor. The square floor is 12 feet long and 12 feet wide.
   a. What is the area in square feet of each tile? (**Hint:** 9 in. = $\frac{3}{4}$ ft)
   b. What is the area of the shop floor?
   c. How many tiles will Ben need to cover this floor?

9. Using square bricks that measure 8 inches on each side, Ellen is building a retaining wall that measures 24 feet long and 4 feet high.
   a. What is the area in square feet of each brick? (**Hint:** 8 in. = $\frac{2}{3}$ ft)
   b. What is the area of the wall?
   c. How many bricks will Ellen need to build this wall?

Two-Step Area Problems

To solve a two-step area problem, divide a complex figure into shapes you're familiar with; then find the area of each shape separately.

···

Find the area of each room pictured below.

10. $A =$ _____
    (**Hint:** Think of the room as being made up of two squares.)

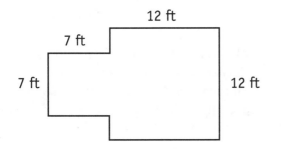

11. $A =$ _____
    (**Hint:** Think of the room as being made up of a square and a rectangle.)

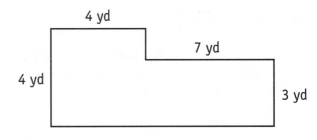

Finding the Area of a Triangle

The area of a triangle is given by the formula $A = \frac{1}{2}bh$, where b stands for *base* and h stands for *height*.

The **base** is one side of the triangle. The **height** is the distance from the base to the vertex of the opposite angle. In the right triangle, the height is one of the sides of the triangle.

$$A = \frac{1}{2}bh$$

> To compute the area of a triangle, multiply $\frac{1}{2}$ times the base times the height.

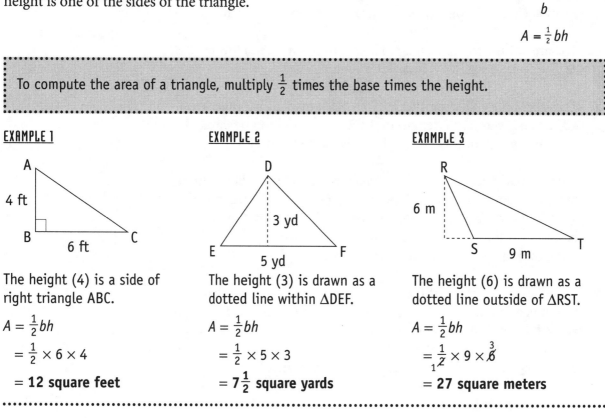

EXAMPLE 1

The height (4) is a side of right triangle ABC.

$A = \frac{1}{2}bh$

$= \frac{1}{2} \times 6 \times 4$

$= $ **12 square feet**

EXAMPLE 2

The height (3) is drawn as a dotted line within ΔDEF.

$A = \frac{1}{2}bh$

$= \frac{1}{2} \times 5 \times 3$

$= 7\frac{1}{2}$ **square yards**

EXAMPLE 3

The height (6) is drawn as a dotted line outside of ΔRST.

$A = \frac{1}{2}bh$

$= \frac{1}{\cancel{2}} \times 9 \times \cancel{6}^{3}$

$= $ **27 square meters**

Find the area of each triangle drawn below.

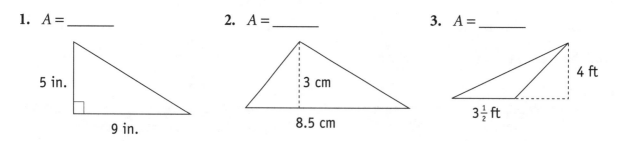

1. $A =$ _____

2. $A =$ _____

3. $A =$ _____

Solve each problem below.

4. Katie's garden is in the shape of a triangle. What is the area of this garden if the base measures 18 yards and the height measures 10 yards?

5. Cal is being charged $1.50 per square foot of plywood. If he buys a triangular piece that measures 8 feet long (the base) by 3 feet high (the height), how much will he be charged?

Finding the Area of a Circle

The area of a circle is given by the formula $A = \pi r^2$, where $\pi \approx 3.14$ or $\pi \approx \frac{22}{7}$ and $r =$ radius.

> To find the area of a circle, multiply π by the square of the radius.

EXAMPLE A circular wading pool has a radius of 14 feet. Find the area of this pool.

Write $\frac{22}{7}$ for π and 14 for r in the area formula.

Use cancellation if possible.

$$A = \pi \times r \times r = \frac{22}{7} \times 14 \times 14 = 616$$

ANSWER: 616 square feet

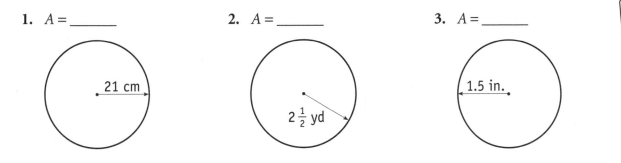

Wading Pool

Find the approximate area of each circle below.

1. $A =$ _____

21 cm

2. $A =$ _____

$2\frac{1}{2}$ yd

3. $A =$ _____

1.5 in.

Solve each problem below.

4. Radio station WPSB broadcasts its signal to all points within a 50-mile radius of the station. What is the approximate area served by this station?

5. A sprinkler waters in a circular pattern, spraying water out 9.5 feet in all directions. To the nearest square foot, what area of lawn does this sprinkler water?

6. A custom-made circular mirror is costing Lou 10¢ per square inch. How much will Lou be charged for a mirror that has a radius of 16 inches?

7. The ratio of the radiuses of two circles is 3 to 1. What is the ratio of their area? (**Hint:** Represent the radius of the larger circle as 3 and the smaller as 1. The number π cancels out when you write the ratio of the two areas.)

Finding the Volume of a Rectangular Solid

Volume is a measure of the space taken up by a solid object or is a measure of the space enclosed by a box or other surface. The most common shape for volume is a **rectangular solid.** It is the familiar shape of boxes, suitcases, freezers, and rooms. The formula for volume of a rectangular solid is $V = lwh$, where l stands for length, w stands for width, and h stands for height. A **cube** is a rectangular solid with the same length, width, and height.

Common **volume units** are the **cubic inch** (cu in.), **cubic foot** (cu ft), and **cubic yard** (cu yd). In the metric system, the most common volume units are the **cubic centimeter** (cm^3) and the **cubic meter** (m^3).

> To find the volume of a rectangular solid, multiply the length times the width times the height.

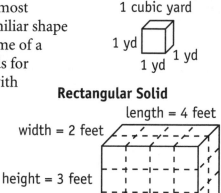

Sample Volume Unit
1 cubic yard

Rectangular Solid

$$V = l \times w \times h$$
$$= 4 \times 2 \times 3$$
$$= \textbf{24 cubic feet}$$

· ·

Find the volume of each rectangular solid.

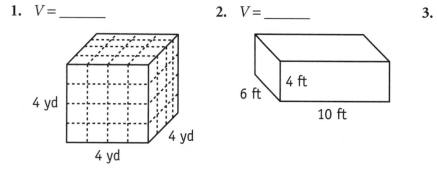

1. $V =$ _____

2. $V =$ _____

3. $V =$ _____

4. A waterbed measures 6 feet long, 5 feet wide, and $\frac{1}{2}$ foot high.

   a. How many cubic feet of water does this waterbed hold?

   b. If water weighs 62 pounds per cubic foot, what is the weight of water in this bed when filled?

5. Billy has two play blocks, each in the shape of a cube. The larger block measures 6 inches on each side. The smaller block measures 2 inches on each side. What is the ratio of the volume of the larger block to the volume of the smaller block?

Finding the Volume of a Cylinder

The volume of a cylinder is given by the formula $V = \pi r^2 h$, where $\pi \approx 3.14$ or $\pi \approx \frac{22}{7}$, r = radius, and h = height.

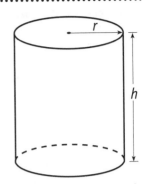

> To find the volume of a cylinder, multiply the area of its bottom (the circle with area πr^2) by its height h.

The top and bottom of a cylinder are equal-size circles.

EXAMPLE A cylindrical water tower has a radius of 21 feet and a height of 15 feet. How many cubic feet of water does this tower hold?

Write $\frac{22}{7}$ for π, 21 for r, and 15 for h.

$V = \pi \times r \times r \times h = \frac{22}{7} \times 21 \times 21 \times 15 = 20{,}790$

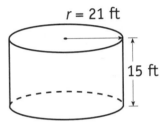

$r = 21$ ft

15 ft

ANSWER: 20,790 cubic feet

Find the *approximate* volume of each cylinder pictured below.

1. $V =$ _____

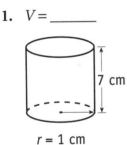

$r = 1$ cm

7 cm

2. $V =$ _____

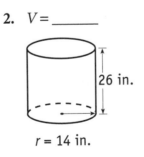

26 in.

$r = 14$ in.

3. $V =$ _____

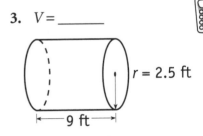

$r = 2.5$ ft

9 ft

Solve each problem below.

4. The cylindrical tank of a gasoline truck is 25 feet long and has a radius of 3.5 feet.

    a. To the nearest cubic foot, what is the volume of this tank?

    b. If 1 cubic foot $\approx$ 7.5 gallons, about how many gallons of gas can the truck carry each load?

5. Two cylinders are the same height. However, the ratio of the radiuses of the cylinders is 3 to 1. Find the ratio of their volumes.
    (**Hint:** Represent the radius of the larger cylinder as 3 and the smaller as 1. The numbers π and h cancel out when you write the ratio of the volumes.)

Geometry Review

Solve the problems below.

1. Francine cut a large circular cake into 12 equal slices. What acute angle is formed by the sides of each slice?

2. What name is given to an angle that measures 124°?

   a. acute   **d.** straight
   b. right   **e.** reflex
   c. obtuse

3. ∠A is *supplementary* to ∠B. What is the sum of ∠A plus ∠B?

4. What is the value of the acute angle that the telephone pole's support cable makes with the ground?

   a. 37°   **c.** 79°   **e.** 114°
   b. 63°   **d.** 92°

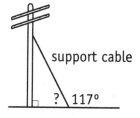

5. What is the value of the acute angle (∠1) that 4th Avenue makes with Oak Street?

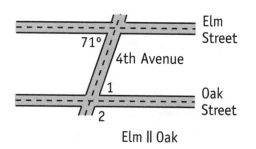

6. What is the value of the obtuse angle (∠2) that Oak Street makes with 4th Avenue?

7. The isosceles triangle at the right has a vertex angle of 50°. What is the measure of the two equal base angles?

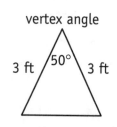

8. ΔABC is similar to ΔXYZ. What is the length of side YZ?

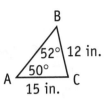

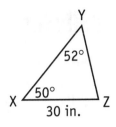

9. On a map that has a scale that reads "1 inch = 250 miles," Jamie measures the distance from Newport to Edison as $2\frac{1}{4}$ inches. What is the approximate distance between these cities?

   a. 400 miles   b. 450 miles   c. 500 miles   d. 550 miles   e. 600 miles

10. Which is the best estimate of the square root of 72?

    a. 6.5   b. 7   c. 7.5   d. 8   e. 8.5

11. The two shorter sides of a right triangle measure 6 feet and 8 feet long. What is the length of the hypotenuse?

12. The rose garden at Central Park is in the shape of a circle. Passing through the center, the distance across the garden is 42 feet. Using $\pi = \frac{22}{7}$, determine the distance around the garden.

13. Matt has moved to a new apartment with an L-shaped living-dining room. Find the area of his living-dining room.

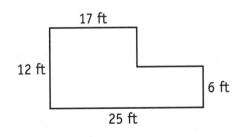

14. How many cubic feet of storage are in a freezer that has inside dimensions of 4 feet by 2.25 feet by 2.5 feet?

15. Using the volume formula $V = \pi r^2 h$, find the volume of the cylindrical drum to the nearest cubic foot.

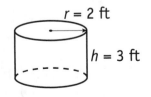

Posttest A

This posttest gives you a chance to check your skill at solving problems covered in this book. Take your time and work each problem carefully. When you finish, check your answers and review any topics on which you need more work.

1. Find the best estimate, to the nearest 10 dollars, for the following purchase: 6 bottles of shampoo at $3.89 per bottle, 8 boxes of detergent at $2.06 per box, and 7 dishtowels at $0.89 per towel.

2. Using the information at the right, estimate to the nearest hundred the total number of packages received at a shipping dock.

Source	Number of Packages
U.S. Postal Service Deliveries	887
Private Parcel Delivery Companies	721
Messenger Deliveries	294

3. Write an inequality using only the symbol < and the numbers −12 and −9.

4. On the number line below, what is the value of $p - q$?

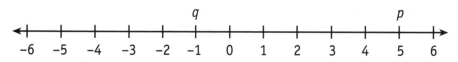

5. On her monthly checking account report, Samantha had a starting balance of $325.50. She saw that during the month she deposited $255.75 to her account and withdrew $170.95 by writing checks. What was the balance at the end of the month?

6. The equation $c = 12 \times \$15.50 - 4 \times \$3.50$ represents the cost c of buying 12 tickets to a baseball game at $15.50 each and using 4 coupons that give a discount of $3.50 each. Find the value of c.

7. The average price of the used cars sold was $8,762. What was that amount rounded to the nearest hundred dollars?

8. Three hundred thirty feet is what fraction of a mile?
   (1 mile = 1,760 yards)

9. A quart container (32 fluid ounces) contains 24 fluid ounces of liquid. Write the fraction in simplest terms that represents the portion of the quart container that is filled.

10. The first four finishers in a 100-meter freestyle swimming event had the following times.

   Darlette   28.307 seconds
   Germaine   28.198 seconds
   Kim   27.075 seconds
   Jann   27.426 seconds

   Which swimmer finished in third place?

11. A carpenter had three pipe wrenches that were the following lengths.

   wrench A: $1\frac{15}{16}$ feet

   wrench B: $1\frac{3}{4}$ feet

   wrench C: $1\frac{7}{8}$ feet

   Which wrench had the middle-size length?

12. To the nearest $\frac{1}{4}$ inch, what is the length of the dowel shown?

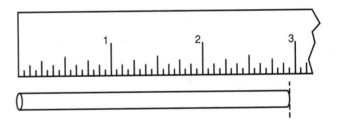

13. A basketball coach told her players that the team had won $\frac{4}{5}$ of their games when they had the lead with 5 minutes left to play. Write $\frac{4}{5}$ as a percent.

14. For a midmorning snack, a company set out a platter with 24 doughnuts. On the platter, 6 doughnuts were filled with lemon custard, 6 doughnuts were chocolate glazed, 4 doughnuts were filled with raspberry custard, and 8 doughnuts were plain. What fraction of the doughnuts were custard-filled? Reduce your answer to lowest terms.

15. In problem 14, is the ratio of plain doughnuts to chocolate doughnuts greater than 1, less than 1, or equal to 1?

16. In an art class with 30 students, only 8 students are men. What is the ratio of men to women?

17. In a recent survey, 2 out of 3 voters supported a proposal to raise money to build a new high school. How many of the 690 people surveyed said they did *not* support the proposal?

18. Shares of stock had the following prices for four weeks.

End of Week 1: $42\frac{1}{2}$

End of Week 2: $45\frac{1}{8}$

End of Week 3: $47\frac{3}{4}$

End of Week 4: $50\frac{3}{8}$

If the trend continues, what will shares cost at the end of Week 5?

 19. One day a taxi driver drove 355 miles on 22.5 gallons of gas. To the nearest tenth of a mile per gallon, what was the taxi's average miles-per-gallon rate for the day?

20. A telephone marketing company has a telephone bill one month of $750 for 20,000 message units. To the nearest tenth of a cent, what was the average charge per message unit?

21. A state has a sales tax rate of 5.5%. Find the amount of sales tax on an item that costs $1,258.

22. The price of a large screen TV was $500 last year. The same TV costs $400 this year. Find the percent of price decrease from last year to this year for the TV.

23. A retired couple bought some vacant land. They made a down payment of $23,400, which represented 30% of the total cost of the land. Find the total cost of the land.

The following data is from a survey about people's movie choices. Use the data for problems 24–26.

Movie Types	No. of People	% of Total
Action	384	32%
Animation	336	28%
Romance	132	
Horror		
Other		
Total		

24. The number of people who chose Horror was half the number of those who chose Action. Find the percent of people who chose Horror.

25. Find the total number of people who took part in the survey.

26. Find the total percent of people who chose either Romance or Other as their favorite movie type.

27. A woman deposited $1,400 into a savings account that paid 5.2% simple interest. Calculate to the nearest cent the amount of interest she will earn at the end of 8 months.

28. A bag contains four marbles: a red one, a blue one, a yellow one, and a green one. You are going to reach into the bag and, without looking, take out one marble at a time. What is the probability that the order you select the marbles will be green, yellow, blue, and then red? Write your answer as a fraction reduced to lowest terms.

29. Which of the three groups of workers at the furniture factory had the greatest decrease from 1995 to 2000?

30. By what percent did the enrollment in the Job Training Program decrease from 1995 to 2000? (Round your answer to the nearest ten percent.)

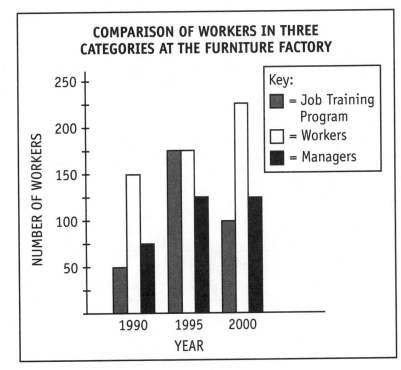

COMPARISON OF WORKERS IN THREE
CATEGORIES AT THE FURNITURE FACTORY

Key:
= Job Training Program
= Workers
= Managers

NUMBER OF WORKERS

YEAR

31. Suppose you spin the spinner 40 times. What is the most likely number of times that the spinner will point to the yellow section?

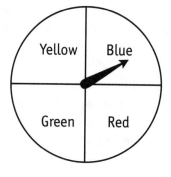

32. What is the probability that both of the next two spins will point to the green section?

33. Write the equation that results when you add 17 to each side of the equation $\frac{2}{3}x - 17 = 4$.

34. A furniture clearance sale announced that all chairs are 60% off. If the sale price of a chair is $29.40, what was the original price of the chair?

35. What is the value of x in the following proportion? $\frac{3}{4} = \frac{12}{x}$

36. Two friends, Victoria and Vanessa, went on a vacation that cost $680. The friends agreed that Vanessa would pay $200 more than Victoria because Vanessa earns more than Victoria. Find the amount that Vanessa paid.

37. At a light bulb assembly line, 3 out of the last 250 light bulbs were defective. At that rate, how many of the next 1,500 light bulbs will be defective?

38. In the equation $y = 5x - 4$, find the value of y when $x = 10$.

39. A long distance telephone company charges a basic monthly fee of $7.80. Then it charges $0.27 each minute of long distance calls made to anywhere in the world. If you make 54 minutes of long distance calls one month, what is the total monthly charge?

40. What are the coordinates of point T?

41. What is the slope of the line through points P and Q?

42. What is the *y*-intercept of the line through points T and S?

43. For the line whose equation is $y = 2x - 3$, the *y*-coordinate of a point on the line is 7. Find the value of the *x*-coordinate.

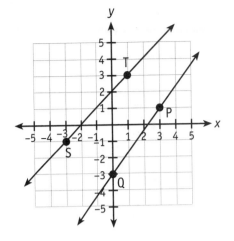

44. A shelf is held up by a support, as shown. What is the value of the acute angle *m* that the shelf makes with the support?

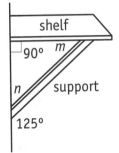

45. A vertical flagpole casts a shadow that is 77 feet long. At the same time, a 3-foot yardstick, held vertically, casts a shadow that is $5\frac{1}{2}$ feet long. Find the height of the flagpole.

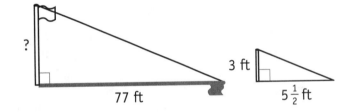

Formulas for problems 46–48:

Pythagorean theorem: $c^2 = a^2 + b^2$

Area of a triangle: $A = \frac{1}{2}bh$

Area of a rectangle: $A = lw$

Volume of a cylinder: $V = \pi r^2 h$

46. The hypotenuse of a right triangle is 9 centimeters and one of the sides is 6 centimeters. Find the length of the other side to the nearest centimeter.

47. A playground is a square that is 20 feet on each side. Part of the playground is covered with concrete and part is covered with grass. Find the area of the part that is covered with grass.

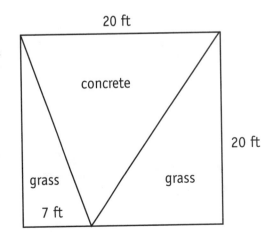

48. A cylindrical tank for a railroad car has a radius of 8 meters. The tank has a height of 22 meters. Find the volume of the tank in cubic meters. (Use 3.14 for π.)

49. A map from an automobile club has a scale of 1 inch = 75 miles. On the map, one straight portion of a highway is $6\frac{1}{4}$ inches long. To the nearest ten miles, what is the actual length of that portion of the highway?

50. An artist has a finished rectangular canvas that is 24 inches high by 32 inches wide. She wants to put a frame all around the canvas, using framing material that is 3 inches wide. What will be the number of square inches of the area of the canvas and its frame?

51. What is the measure of an angle that is one sixth of a circle?

52. Find the solution of the equation $3x - 7 = 8$.

53. A radio station sponsored a preview for a movie about the 1849 Gold Rush in California. At the preview, all 160 seats in the theater were filled. The radio station had placed special tickets under 20 of the seats in the theater. Find the probability of winning a special ticket for each person at the preview. Write your answer as a fraction in lowest terms.

54. If a car averages 22.5 miles per gallon for highway driving, how many gallons of gasoline will it use to drive 315 miles on a highway?

55. In this diagram, line *V* is parallel to line *W.*
How many angles in the diagram are the same size as ∠1? (Do not include ∠1 in your count.)

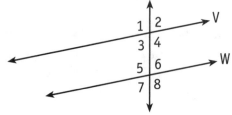

56. During a training program, a National Guard vehicle drove 3 kilometers north and 2 kilometers east. Then it drove 1 kilometer south, 6 kilometers west, and 2 kilometers south. How many kilometers was it from its starting position? (Make a drawing to see where the vehicle is.)

Posttest A Prescriptions

Circle the number of any problem that you miss. If you missed one or less in each group below, go on to Using Number Power. If you missed more than one problem in any group, review the pages in this book or refer to these practice pages in other materials from Contemporary Books.

PROBLEM NUMBERS	PRESCRIPTION MATERIALS	PRACTICE PAGES
1, 2, 3, 4, 5, 7	number sense	10–27
8, 9, 11, 12, 13, 14, 15, 16, 18	fractions	60–67, 83–107
10, 19, 20, 54	decimals	114–123
21, 22, 23, 27, 30, 34	percents	70–73, 132–153
	Math Exercises: Fractions/ Decimals/Percents	3–29 in each book
	Breakthroughs in Math, Book 2	34–141
	Math Skills That Work, Book Two	32–139
	Real Numbers: Fractions and Percents	1–68
	Real Numbers: Whole Numbers and Decimals	37–69
	Critical Thinking with Math	66–105
24, 25, 26, 28, 29, 31, 32, 53	data analysis, probability	158–173
	Math Exercises: Data Analysis & Probability	3–29
	Math Skills That Work, Book Two	160–179
	Real Numbers: Table, Graphs, & Data Interpretation	1–65
6, 17, 33, 35, 36, 37, 38, 39, 40, 41, 42, 43, 52	algebra	176–201
	Math Exercises: Pre-Algebra	3–29
	Math Exercises: Algebra	3–29
	Real Numbers: Algebra Basics	1–67
44, 45, 46, 47, 48, 49, 50, 51, 55, 56	geometry	204–225
	Math Exercises: Geometry	3–29
	Real Numbers: Geometry Basics	1–68

For further practice:

Math Solutions (software)

Posttest B

This text has a multiple-choice format much like many standardized tests. Take your time and work each problem carefully. Circle the correct answer to each problem.

1. For her salon, Gina bought 11 bottles of shampoo on sale for $2.79 per bottle. Which of the following is the *best estimate* of the change Gina will receive if she pays with two $20 bills?

   a. $5   **b.** $10   **c.** $15   **d.** $20   **e.** $25

2. *Estimate* the total number of customers who visited Family Burgers during the four days shown.

   a. 2,000   **d.** 3,200
   b. 2,400   **e.** 3,600
   c. 2,800

Family Burgers Customer Count

Day	Customers
Monday	689
Tuesday	714
Wednesday	733
Thursday	673

3. Which of the following inequalities is *not* true?

   a. $-5 > -3$ **d.** $3 > -2$
   b. $0 > -2$ **e.** $7 > 0$
   c. $-4 < -1$

4. Using values represented on the number line, what is $x - y$?

   a. -5 **d.** 1
   b. -1 **e.** 5
   c. 0

$$y = -2 \qquad x = 3$$
$$-4 \quad -3 \quad -2 \quad -1 \quad 0 \quad 1 \quad 2 \quad 3 \quad 4$$

5. Playing on the game show "You Keep the Money," Todd Jackson had a balance of $-$350$ in winnings after round 2. In round 3, though, Todd won $750. What was Todd's balance after round 3?

   a. $400   **b.** $600   **c.** $800   **d.** $1,000   **e.** $1,100

6. Bruce purchased a case of car oil that was marked down $2.00 below its normal price of $12.40. Bruce was also given a rebate coupon for $1.50. Which equation represents the cost (c) to Bruce for the case of oil after he received the rebate?

   a. $c = \$12.40 - \$2.00 + \$1.50$ **c.** $c = \$12.40 + \$2.00 - \$1.50$
   b. $c = \$12.40 - \$2.00 - \$1.50$ **d.** $c = \$12.40 + \$2.00 + \$1.50$

7. The average price of a condominium in a new development is $119,231. What is this amount rounded to the nearest thousand dollars?

   a. $100,000   b. $110,000   c. $119,000   d. $120,000   e. $125,000

8. Two hundred twenty yards is what fraction of a mile?
   (1 mile = 5,280 feet)

   a. $\frac{1}{24}$ b. $\frac{1}{16}$ c. $\frac{1}{10}$ d. $\frac{1}{8}$ e. $\frac{1}{4}$

9. A cup (8 fluid ounces) is filled with 6 fluid ounces of milk. Which of the following does *not* represent the part of the cup that is filled?

   a. $\frac{6}{8}$ b. 75%   c. $\frac{3}{4}$ d. 0.75   e. 68%

10. Listing the winner first, in what order did the four runners finish?

    a. Young-Soo, Pat, Kenny, Carlo
    b. Pat, Young-Soo, Carlo, Kenny
    c. Carlo, Kenny, Young-Soo, Pat
    d. Kenny, Carlo, Young-Soo, Pat
    e. Young-Soo, Kenny, Carlo, Pat

    200-Meter Race Times

    Carlo: 24.07 sec
    Young-Soo: 23.8 sec
    Kenny: 24.1 sec
    Pat: 23.795 sec

11. Andre's boss asked him to arrange the following bolts in a storage drawer according to length, placing the shortest bolt near the front.

    $1\frac{3}{4}$ in. $1\frac{5}{8}$ in. $1\frac{11}{16}$ in.

    Bolt A   Bolt B   Bolt C

    Listing the shortest bolt first, what is the correct order of bolts?

    a. A, B, C   b. A, C, B   c. B, A, C   d. B, C, A   e. C, B, A

12. To the nearest $\frac{1}{8}$ inch, what is the width of the board shown?

    a. 2 in.   d. $2\frac{3}{8}$ in.

    b. $2\frac{1}{8}$ in.   e. $2\frac{1}{2}$ in.

    c. $2\frac{1}{4}$ in.

13. During winter-league play, Laurie had scores of 150 or more in 73% of the games she bowled. In about what fraction of her games did Laurie score 150 or more?

    a. $\frac{1}{2}$ b. $\frac{5}{8}$ c. $\frac{2}{3}$ d. $\frac{3}{4}$ e. $\frac{7}{8}$

For problems 14 and 15, refer to the following information.

Andrew places the following ingredients in an 8-cup mixing bowl: $\frac{7}{8}$ cup of white flour, $1\frac{3}{4}$ cups of whole-wheat flour, $\frac{1}{2}$ cup of wheat bran, and $\frac{7}{8}$ cup of dark rye flour.

14. At this point, what fraction of the mixing bowl has Andrew filled?

    a. $\frac{1}{8}$ b. $\frac{1}{4}$ c. $\frac{3}{8}$ d. $\frac{1}{2}$ e. $\frac{5}{8}$

15. In Andrew's mixture, what is the ratio of the amount of whole-wheat flour to the amount of white flour?

    a. $\frac{1}{2}$ b. $\frac{4}{7}$ c. $\frac{1}{1}$ d. $\frac{3}{2}$ e. $\frac{2}{1}$

16. A 32-fluid-ounce punch mixture contains orange juice and lime soda. The mixture contains just 12 fluid ounces of lime soda. In this mixture, what is the ratio of lime soda to orange juice?

    a. $\frac{3}{5}$ b. $\frac{3}{8}$ c. $\frac{5}{8}$ d. $\frac{5}{3}$ e. $\frac{8}{3}$

17. Working 5 days a week, 8 hours a day, Ravi earns $288 a week. Next week, though, Ravi plans to work only 24 hours. For this work, he'll receive this same hourly pay. Which expression below tells how much Ravi will earn next week?

    a. $\frac{3}{5}$ ($288)   c. $\frac{2}{3}$ ($288)   e. $288 - \frac{3}{5}$ ($288)

    b. $\frac{2}{5}$ ($288)   d. $288 - \frac{2}{3}$ ($288)

18. The average monthly value of NMP Company stock for a 3-month period is shown. If this trend continues, what will be NMP's average monthly stock value for June?

    a. $35\frac{3}{4}$ d. $35\frac{3}{8}$

    b. $35\frac{5}{8}$ e. $35\frac{1}{4}$

    c. $35\frac{1}{2}$

NMP Average Monthly Stock Value		
March	April	May
$38\frac{1}{4}$	$37\frac{3}{8}$	$36\frac{1}{2}$

19. During the week, Lanna drove her car a total of 268 miles on 13.8 gallons of gas. To the nearest tenth of a mile per gallon, how many miles per gallon is Lanna's car averaging?

    a. 19.4   b. 19.9   c. 20.3   d. 20.8   e. 21.2

20. The Changs' heating bill for January is $329. During January, they used 4,000 kilowatt-hours of electric power. To the nearest *tenth of a cent*, what are the Changs being charged per kilowatt-hour?

    a. $0.80   b. $0.08   c. $0.082   d. $0.083   e. $8.3

21. Aurora receives a 6% real estate commission for each house she sells. What will be Aurora's commission on a house that sells for $84,000?

    a. $504   b. $1,540   c. $4,840   d. $5,040   e. $50,400

22. A washing machine that sold for $319.99 last year is priced at $399.89 this year. Which expression below gives the best *estimate* of the *percent of price increase* of this washing machine during the past year?

    a. $\frac{\$320}{\$400} \times 100\%$ c. $\frac{\$400 - \$320}{\$320} \times 100\%$ e. $\frac{\$320 + \$400}{\$320} \times 100\%$

    b. $\frac{\$400}{\$320} \times 100\%$ d. $\frac{\$400 - \$320}{\$400} \times 100\%$

23. To purchase a new pickup truck, Chen must make a 20% down payment. If he is required to put down $3,250, what is the price of the truck?

    a. $12,920   b. $13,540   c. $14,760   d. $15,840   e. $16,250

24. In a public opinion poll, 119 of the 548 people surveyed said they *do not* plan to buy an American car as their next car. Which expression gives the percent of those surveyed who *do* plan to buy an American car as their next car?

    a. $\frac{119}{548} \times 100\%$ c. $\frac{548 - 119}{119} \times 100\%$ e. $\frac{119}{548 - 119} \times 100\%$

    b. $\frac{548}{119} \times 100\%$ d. $\frac{548 - 119}{548} \times 100\%$

25. What total percent of ingredients in a 5-pound box of Goodgrow Plant Food is not written on the label shown?

    a. 30%   d. 60%
    b. 40%   e. 70%
    c. 50%

26. In a 5-pound box of Goodgrow Plant Food, *about* what total weight is made up of phosphorus and calcium? (1 pound = 16 ounces)

    a. 9 ounces   d. 1 pound 11 ounces
    b. 13 ounces   e. 2 pounds 7 ounces
    c. 1 pound 4 ounces

Goodgrow Plant Food
Net Wt. 5 lb

Ingredients	
Phosphorus	12%
Potassium	8%
Nitrogen	4%
Calcium	4%
Magnesium	2%

27. Suppose you deposit $750 in a savings account that pays 4.9% simple interest. Which expression is the best *estimate* of the total in your account at the end of 2 years 1 month?

a. $750 + ($750 × 0.05 × 2)
b. $750 − ($750 × 0.5 × 2)
c. $750 + ($750 × 0.5 × 2)
d. $750 − ($750 × 0.05 × 2)
e. $750 × 0.05 × 2

28. In which town and decade did average prices increase the most?

a. Fernwood 1980 to 1990
b. Oak Grove 1980 to 1990
c. Monroe 1980 to 1990
d. Fernwood 1990 to 2000
e. Oak Grove 1990 to 2000

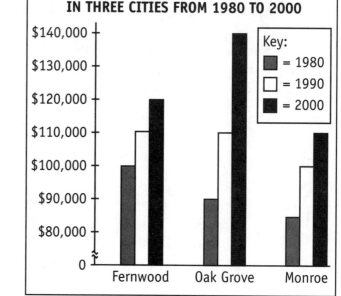

AVERAGE PRICE OF NEW CONDOMINIUMS IN THREE CITIES FROM 1980 TO 2000

Key:
■ = 1980
□ = 1990
■ = 2000

Fernwood   Oak Grove   Monroe

29. By *about* what percent did average prices increase in Monroe between 1990 and 2000?

a. 10%
b. 20%
c. 30%
d. 40%
e. 50%

30. The three numbers of Ruth's combination lock are 8, 19, and 31, though she can't remember which order they're in! What's the probability that the correct combination is 19, 31, 8?

a. $\frac{1}{9}$
b. $\frac{1}{6}$
c. $\frac{1}{5}$
d. $\frac{1}{3}$
e. $\frac{1}{2}$

31. Of the next 80 students who register at Lewis College, how many most likely will be from Asia?

a. 12
b. 15
c. 18
d. 21
e. 24

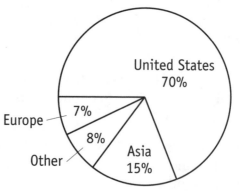

LEWIS COLLEGE PLACE OF ORIGIN OF STUDENTS

United States 70%

Europe — 7%

Other — 8%

Asia 15%

32. At a Memorial Day sale, Lita bought a sweater marked 20% off. If Lita paid $24 for the sweater, which equation below can be solved to find the original price (p)?

    a. $p = \frac{1}{5}(\$24)$

    b. $p = p + \frac{1}{5}(\$24)$

    c. $p - \frac{1}{5}p + \$24 = 0$

    d. $p - \frac{1}{5}p = \$24$

    e. $p + \frac{1}{5}p = \$24$

33. What equation shows the equal cross products in the proportion $\frac{2}{3} = \frac{8}{x}$?

    a. $3x = 2(8)$ b. $2(3) = 8x$ c. $2x = 3(8)$ d. $\frac{2}{3} = \frac{8}{x}$ e. $\frac{x}{8} = \frac{3}{2}$

34. Alex and Joel have a yard-care service. Because Alex owns the truck and lawn-care equipment, he earns $25 more each day than Joel does. On Saturday the two men earned a total of $165. Which equation below can be solved to determine Joel's share (x) of Saturday's earnings?

    a. $x - x + \$25 = \$165$

    b. $x + x + \$25 = \$165$

    c. $x + x = \$165 + \$25$

    d. $x - \$25 = \$165$

    e. $x + \$25 = \$165$

35. Seven out of the first 230 VCRs tested were found to have a tracking problem. Seven hundred seventy more VCRs are going to be tested. Which proportion can be used to estimate the number (n) of these remaining VCRs that will likely have a similar problem?

    a. $\frac{7}{230} = \frac{1000}{n}$

    b. $\frac{7}{230} = \frac{n}{1000}$

    c. $\frac{7}{230} = \frac{770}{n}$

    d. $\frac{7}{230} = \frac{n}{770}$

    e. $\frac{230}{770} = \frac{n}{1000}$

36. In the equation $y = 3x - 2$, what is the value of y when $x = 4$?

    a. 2   b. 6   c. 10   d. 12   e. 14

37. Emerald Car Rental charges $19 a day plus $0.15 per mile for rental of its compact cars. Letting t stand for total cost and n for number of days of rental, which equation below represents the total cost of renting a compact car from Emerald?

    a. $t = \$19 + \$0.15n$

    b. $t = \$19 - \$0.15n$

    c. $t = \$19n + \$0.15$

    d. $t = \$19n - \$0.15$

    e. not enough information given

38. Which of the lettered points has the coordinates (4, –2)?

   a. A   **c.** C   **e.** E
   b. B   **d.** D

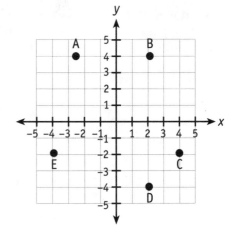

39. When graphed, which of the following equations passes through the point (3, 4)?

   a. $y = x - 3$ **d.** $y = \frac{1}{3}x + 1$

   b. $y = x + 1$ **e.** $y = 2x - 1$

   c. $y = \frac{2}{3}x + 3$

40. What is the slope of line L?

   a. -1 **c.** 1 **e.** $\frac{3}{2}$
   b. 0 **d.** $\frac{2}{3}$

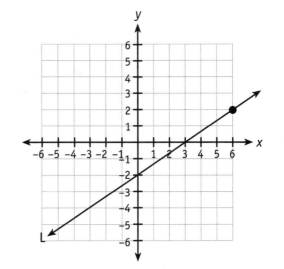

41. What is the *y*-intercept of line L?

   a. $(0, -2)$ **c.** $(-2, 0)$ **e.** $(3, 0)$
   b. $(-2, 3)$ **d.** $(3, -2)$

42. A support cable is used to help hold a radio broadcasting antenna in a vertical position. What is the value of the acute angle *y* that the cable makes with the antenna?

   a. 25°   **c.** 75°   **e.** 90°
   b. 65°   **d.** 80°

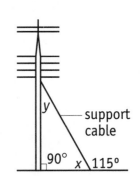

43. Gerald is standing near a telephone pole. Both he and the pole cast shadows on the level ground. Which proportion is the correct one to use to determine the height (*h*) of the pole?

   a. $\frac{3}{5} = \frac{160}{h}$ **d.** $\frac{h}{8} = \frac{160}{5}$

   b. $\frac{h}{5} = \frac{160}{8}$ **e.** not enough information given

   c. $\frac{3}{5} = \frac{h}{160}$

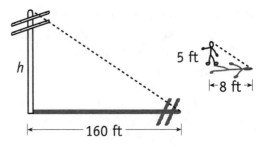

Formulas for problems 44–46:

Pythagorean theorem $c^2 = a^2 + b^2$

Area of a triangle: $A = \frac{1}{2}bh$

Area of a rectangle: $A = lw$

Volume of a cylinder: $V = \pi r^2 h$

44. The two shorter sides of a right triangle measure 5 feet and 6 feet. What is the best *estimate* of the length of the hypotenuse?

   a. 4 ft   **b.** 5 ft   **c.** 6 ft   **d.** 7 ft   **e.** 8 ft

45. The garden space in Shanda's backyard is in the shape of a triangle. The rest of the yard is planted in grass. How many square yards of grass are in this yard?

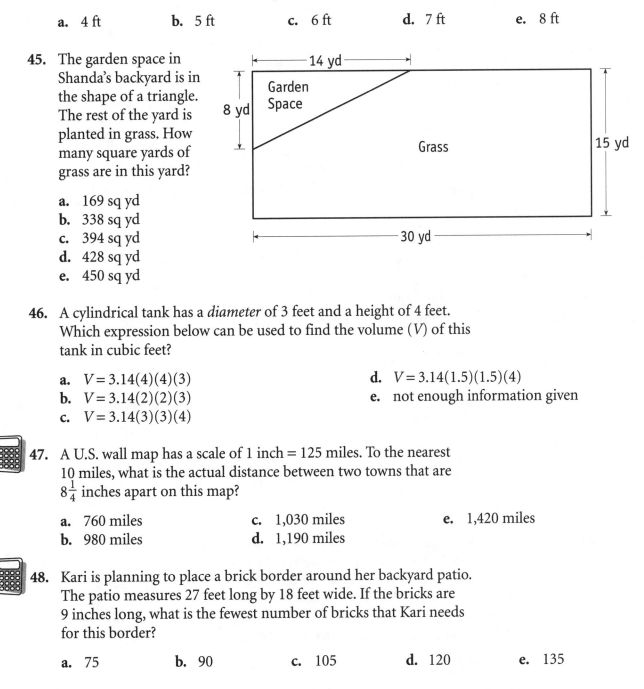

   a. 169 sq yd
   b. 338 sq yd
   c. 394 sq yd
   d. 428 sq yd
   e. 450 sq yd

46. A cylindrical tank has a *diameter* of 3 feet and a height of 4 feet. Which expression below can be used to find the volume (V) of this tank in cubic feet?

   a. $V = 3.14(4)(4)(3)$ **d.** $V = 3.14(1.5)(1.5)(4)$
   b. $V = 3.14(2)(2)(3)$ **e.** not enough information given
   c. $V = 3.14(3)(3)(4)$

47. A U.S. wall map has a scale of 1 inch = 125 miles. To the nearest 10 miles, what is the actual distance between two towns that are $8\frac{1}{4}$ inches apart on this map?

   a. 760 miles   **c.** 1,030 miles   **e.** 1,420 miles
   b. 980 miles   **d.** 1,190 miles

48. Kari is planning to place a brick border around her backyard patio. The patio measures 27 feet long by 18 feet wide. If the bricks are 9 inches long, what is the fewest number of bricks that Kari needs for this border?

   a. 75   **b.** 90   **c.** 105   **d.** 120   **e.** 135

49. The whole pie is cut into 12 equal pieces. What angle do the sides of each piece make?

    a. 36°   d. 50°
    b. 90°   e. 30°
    c. 45°

50. Which equation does *not* have the same solution as the others?

    a. $2x + 2 = x + 4$ d. $2x = 14 - 6$
    b. $x + 8 = 12$ e. $2x - 3 = x + 1$
    c. $x \div 2 = 2$

51. As part of a fundraising effort, Kim's class sold 225 raffle tickets for $4.00 each. The winner takes home a gas barbecue. Kim's parents bought three tickets. What is the probability that Kim's family will win the barbecue?

    a. $\frac{1}{225}$ b. $\frac{1}{75}$ c. $\frac{2}{75}$ d. $\frac{4}{225}$ e. $\frac{12}{225}$

52. Jan's car used $9\frac{1}{2}$ gallons of gasoline to drive between Rockford and Glendale. What more do you need to know to be able to determine what mileage Jan's car got on this trip?

    a. the price Jan paid per gallon of gasoline
    b. the highway mileage rating of Jan's car when new
    c. the length of time it took Jan to make the trip
    d. the distance between Rockford and Glendale
    e. the city mileage rating of Jan's car when new

53. Which of the following do you need to know to find the value of $\angle 1$?

    A. the sum of $\angle 2$ and $\angle 3$
    B. that line L is parallel to line M
    C. the value of $\angle 3$

    a. A only   d. A and B
    b. B only   e. B and C
    c. C only

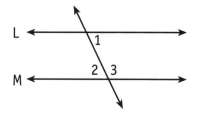

54. Starting at the library on Jenson Avenue, Na walked 5 blocks east before turning south on 21st Street. She then walked 8 blocks, turned west onto Hilyard Avenue, and walked 12 more blocks to 33rd Street. How many blocks is Na from the library now if she can walk north on 33rd Street to Jenson and then turn east to go to the library? (Make a drawing to find out where Na is now.)

    a. 12   b. 15   c. 18   d. 25   e. 31

Posttest B Chart

If you missed more than one problem in any group below, review the practice pages for those problems. Then redo the problems you got wrong before going on to Using Number Power.

PROBLEM NUMBERS	SKILL AREA	PRACTICE PAGES
1, 2, 3, 4, 5, 7	number sense	10–27
8, 9, 11, 12, 14, 15, 16, 17, 18, 52	fractions	60–67, 82–105
10, 19, 20	decimals	114–123
13, 21, 22, 23, 24, 25, 26, 27, 32	percents	70–73, 132–153
28, 29, 30, 31, 51	data analysis, probability	158–173
6, 33, 34, 35, 36, 37, 38, 39, 40, 41, 50	algebra	176–201
42, 43, 44, 45, 46, 47, 48, 49, 53, 54	geometry	204–225

Using
Number
Power

Using Drawings

Sometimes a word problem can seem confusing. One technique that can be helpful for a confusing problem is to use a drawing to organize the information. The saying "A picture is worth 1,000 words" is especially true for math problems!

EXAMPLE   Marcella, Tony, and William carpool to QRS Electronics, where they work. All three live on 46th Street, east of QRS. Marcella lives 13.9 miles farther from work than Tony. William lives 11.5 miles closer to work than Marcella. Tony lives 12.3 miles from QRS. How far does William live from QRS?

STEP 1   Make a drawing, starting with the easiest fact to represent. *Tony lives 12.3 miles from QRS Electronics.*

|←———— 12.3 ————→|

●————————————————●————————————————————
QRS Electronics   Tony

STEP 2   Represent the other two facts on this drawing. *Marcella lives 13.9 miles farther from work than Tony. William lives 11.5 miles closer to work than Marcella.*

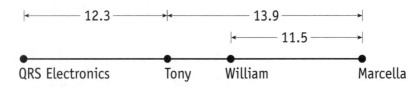

STEP 3   Use the drawing to solve the problem.
  a. Determine how far Marcella lives from QRS Electronics.
    12.3 + 13.9 = 26.2 miles
  b. William lives 11.5 miles closer than Marcella.
    26.2 − 11.5 = 14.7 miles

ANSWER: William lives 14.7 miles from QRS Electronics.

For each problem, use a drawing to organize the facts; then solve.

1.   Karin, Shauna, and Amanda all live on Walnut Road west of the library. Karin lives 3.4 miles farther from the library than Shauna. Amanda lives 1.6 miles closer to the library than Karin. Shauna lives 5.8 miles from the library. How far does Amanda live from the library?

Drawing

2. An airplane is flying at an altitude of 35,000 feet above sea level. A cloud, 18,500 feet below the airplane, is passing over Mount Scott, the summit of which is 9,500 feet above sea level. How far is the cloud above the summit of Mount Scott?

    Drawing

3. Stanley knows a secret and told it to Jeff. During the day, Jeff told four friends the secret. Each of these friends told three other friends. How many people, not counting Stanley, know the secret now?

    Drawing

4. Between 7:00 A.M. and 10:30 A.M., the temperature rose 24°F. However, the temperature rose only 9° between 9:00 A.M. and 10:30 A.M. What was the temperature at 9:00 A.M. if the temperature at 7:00 A.M. was 47°F?

    Drawing

5. From his home on Filmore Street, Tuan walked 1.4 miles directly east. He then turned onto 34th Street and walked 0.6 miles directly south. At that point he turned onto Harrison Street and walked 0.9 miles directly west. How far is Tuan now from home if he can get there by walking directly north to Filmore Street and then turning west to go home?

    Drawing

6. Bobbie left Seattle at 8:00 A.M. and drove south toward Eugene at a speed of 60 miles per hour. Diane left Eugene at 9:00 A.M. and drove north toward Seattle at a speed of 50 miles per hour. If Seattle is 300 miles from Eugene, how far apart were Bobbie and Diane at 10:00 A.M.? (**Hint:** As a first step, determine how far each person had driven by 10:00 A.M.)

    Drawing

Working Backward

In some problems, you are given an end value and asked to find a previous value. For problems of this type, you'll often find that *working backward* is the best approach. Below are two examples.

EXAMPLE 1 After getting up, Jerry takes 45 minutes to eat and get ready for work. He then drives 10 minutes to Frank's, where he meets Frank, Olivia, and Jesse. Together they carpool to their office, a drive that takes 35 minutes. They always allow 30 minutes of extra time to have coffee at the office cafeteria before work.

a. What time must they leave Frank's if they need to be at work by 9:00 A.M.?

Work backward, using the stated facts to find the time of each event.

Time they arrive at coffee shop: 9:00 A.M. − 30 minutes = 8:30 A.M.
Time they leave Frank's: 8:30 A.M. − 35 minutes = 7:55 A.M.

ANSWER: They must leave Frank's by 7:55 A.M.

b. What time must Jerry get up if he wants to be at work by 9:00 A.M.?

Work backward from the answer in part a.

Time Jerry leaves for Frank's: 7:55 A.M. − 10 minutes = 7:45 A.M.
Time Jerry must get up: 7:45 A.M. − 45 minutes = 7:00 A.M.

ANSWER: Jerry must get up by 7:00 A.M.

EXAMPLE 2 The cost of a hotel suite at a vacation resort has risen drastically between 1980 and today. Between 1980 and 1990, the cost doubled. Between 1990 and 1992, it increased another $60 per week. Today's cost of $850 per week is $340 higher than it was in 1992. How much was a hotel suite in 1980?

Work backward, using the stated facts to figure the weekly rent for each given year.

a. Cost today: $850
b. Cost in 1992: $850 − $340 = $510
c. Cost in 1990: $510 − $60 = $450
d. Cost in 1980: $450 ÷ 2 = $225

ANSWER: The cost of a hotel suite in 1980 was $225.

Work backward to solve each problem.

1. Mr. Johnson kept 12 of the cookies he made and placed them in a container for his family. He took the rest of the cookies to his daughter's class. The teacher, Mr. Swenson, gave half of these cookies to his class and the other half to Mrs. Blake's class. Mrs. Blake had 32 cookies to share with her students.

   a. How many cookies did Mr. Swenson and Mrs. Blake share?

   b. How many cookies did Mr. Johnson bake?

2. After she wakes up, it takes Jillian 1 hour 30 minutes to eat, get her daughter off to school, and get herself ready to leave for work. She then has a 10-minute walk to the bus station and a 15-minute bus ride to the stop close to her work. From this stop, it is a 5-minute walk to work.

   a. To be at work by 9:30 A.M., what time must Jillian catch the bus?

   b. What time must Jillian get out of bed if she wants to be to work by 9:30 A.M.?

3. Jackson used $\frac{1}{2}$ of a full box of nails while working on the upstairs of a new house. Loretta, who was working on the main floor, took $\frac{1}{2}$ of the nails left in the box. When Tom went to get nails, he found only 1 pound of nails remaining.

   a. How many pounds of nails were in the box just before Loretta took some out?

   b. How many pounds of nails were in a full box?

4. Tuesday's low temperature was 12 degrees higher than Monday's low. Wednesday's low of 42°F was 9 degrees colder than Tuesday's low. Use these facts to determine Monday's low temperature reading.

5. The average starting hourly wage at Jenson Electronics has gone up substantially between 1960 and today. Between 1960 and 1965, the average starting wage increased by $1.50 per hour. Between 1965 and 1990, the starting wage doubled. Between 1990 and today, the starting wage increased by another $1.75 per hour. If today's starting wage is $11.65, what was the average starting wage in 1960?

Working with Rates

Closely related to ratios are **rates.** Like a ratio, a rate is a comparison of two numbers. A rate, though, always compares one number of units to a different, single unit.

Rates are usually expressed with the word *per. Per* means "for each."

You may already be familiar with several commonly used rates.

speed $= \frac{\text{miles}}{\text{hour}}$ (miles per hour)   **mileage** $= \frac{\text{miles}}{\text{gallon}}$ (miles per gallon)

unit price $= \frac{\text{dollars}}{\text{pound}}$ (dollars per pound)   **heart rate** $= \frac{\text{beats}}{\text{minute}}$ (beats per minute)

Often you are asked to find a rate from a given ratio.

EXAMPLE 1   Aurora drove 171 miles in 3 hours. On average, how many miles did Aurora drive each hour?

The question is asking you to find Aurora's average rate of *speed*—the average number of miles she drove in 1 hour.

To find this rate, follow these steps.

STEP 1   Write a ratio comparing miles to hours.

$$\frac{\text{miles}}{\text{hours}} = \frac{171 \text{ miles}}{3 \text{ hours}}$$

STEP 2   Find the rate by dividing the numerator (171) by the denominator (3).

$$\frac{171 \text{ miles} \div 3}{3 \text{ hours} \div 3} = \frac{57 \text{ miles}}{1 \text{ hour}}$$

(**Remember:** A rate in fraction form has a denominator of 1. The rate $\frac{57 \text{ miles}}{1 \text{ hour}}$ can also be expressed as 57 miles per hour *or* 57 mph.)

ANSWER: The average speed is 57 miles per hour *or* 57 mph.

EXAMPLE 2   At Hong's Oriental Market, Li bought 4 pounds of winter melon for $11.56. What price was Li charged per pound?

This question asks you to find the *unit price* that Li is paying—the amount she is paying for 1 pound of winter melon.

To find this rate, follow these steps.

STEP 1   Write a ratio comparing price to pounds.

$$\frac{\text{price}}{\text{pounds}} = \frac{\$11.56}{4 \text{ pounds}}$$

STEP 2   Find the unit price by dividing $11.56 by 4.

$$\frac{\$11.56 \div 4}{4 \text{ pounds} \div 4} = \frac{\$2.89}{1 \text{ pound}}$$

ANSWER: The unit price is $2.89 per pound *or* $2.89/lb.

Find the rates in the following problems.

1. Over a 5-minute period, Ben counted his heartbeats. He counted 380 beats. At what rate (beats per minute) is Ben's heart beating?

2. A high-speed train in Europe traveled 570 miles in 3 hours. What was the average speed of this train during the trip?

3. Valley Car Rental has a 5-day special car rental package for $165. What is the daily rate that Valley is charging on this package?

4. Rosie rode her bike from Bay City to Evans, a distance of 72 miles. The trip took $4\frac{1}{2}$ hours. What was Rosie's average speed for this distance?

5. The results of a gas mileage (miles per gallon) test performed on four cars is shown. Compute the gas mileage of each car. Write your answers in the table.

Car	Miles Driven	Gas Used	Gas Mileage
#1	144	6 gal	
#2	195	$7\frac{1}{2}$ gal	
#3	65	$3\frac{1}{4}$ gal	
#4	78	$4\frac{1}{3}$ gal	

6. During a rainstorm, $1\frac{1}{4}$ inches of rain fell in Lakeview in a $4\frac{1}{4}$-hour period. What was the average rate of rainfall in inches per hour?

7. Marlena earned $40.25 for $3\frac{1}{2}$ hours of overtime work on Saturday. Determine Marlena's overtime pay rate.

8. Abdulla walked $8\frac{3}{4}$ miles in $2\frac{1}{2}$ hours. Expressing your answer in miles per hour, determine Abdulla's average walking rate.

9. The Community Wading Pool was drained so that repairs could be made. It took 3 hours 20 minutes to drain the pool's 15,000 gallons of water. What was the drainage rate in gallons per hour? (**Hint:** Write 20 minutes as a fraction of an hour.)

10. To the nearest cent, compute the price per ounce of each size of soft drink shown below.

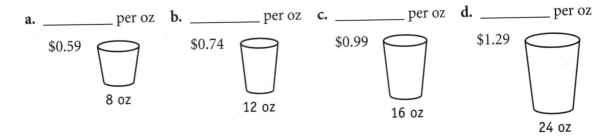

a. _____ per oz   b. _____ per oz   c. _____ per oz   d. _____ per oz

$0.59   8 oz $0.74   12 oz $0.99   16 oz $1.29   24 oz

Working with Similar Figures

Proportions can be used to solve problems involving other types of similar figures. Any two figures are similar if their corresponding angles are equal and their corresponding sides are in proportion.

Enlarging or Reducing a Picture

You produce a similar figure when you enlarge or reduce a photograph or drawing.

EXAMPLE 1   Marsha wants to enlarge a photo that measures 3 inches high by 5 inches wide. She wants the enlargement to be 12 inches wide. What will be the height of the enlargement?

h

12 in.
Enlargement

STEP 1   Letting h be the height of the copy, write a proportion as follows.

$$\frac{\text{height of copy}}{\text{height of original}} = \frac{\text{width of copy}}{\text{width of original}}$$

$$\frac{h}{3} = \frac{12}{5}$$

STEP 2   Solve for h.

$$5h = 36$$

$$\frac{5h}{5} = \frac{36}{5}$$

$$h = 7\frac{1}{5} \text{ inches}$$

3 in.

5 in.
Original

ANSWER: The enlargement will be $7\frac{1}{5}$ inches in height.

Working with a Scale Drawing

Scale drawings, such as blueprints and maps, are drawings that are *similar* to the represented object. The **scale** tells the ratio of the corresponding dimensions. For example, if a scale is "1 inch = 24 inches," the ratio of the drawing to the actual object is 1 to 24.

EXAMPLE 2   Hal is looking at his house blueprint. The scale reads "1 inch = 18 inches." What is the actual width of a hallway that measures $2\frac{1}{4}$ inches on the blueprint?

To find the width of the hallway, multiply the drawing width by 18.

Hallway width = $18 \times 2\frac{1}{4} = 40\frac{1}{2}$ inches

ANSWER: $40\frac{1}{2}$ inches

EXAMPLE 3   A map of the United States has a scale that reads "1 inch = 250 miles." What is the distance between the two cities whose map distance is $6\frac{1}{2}$ inches?

To find the distance, multiply the map distance by 250.

Distance = $250 \times 6\frac{1}{2} = 1,625$ miles

ANSWER: 1,625 miles

Solve each problem.

1. As shown, Connie is going to have a photograph enlarged. The larger copy is to be 18 inches wide. What will be the height of this copy?

4 in.

6 in.
Original Photo

?

18 in.
Enlargement

2. On a building blueprint, a room measures $6\frac{1}{2}$ inches wide by 9 inches long. The blueprint scale reads "1 inch = 2 feet." What are the actual dimension of this room?

   width:_____ length:_____

3. As shown, a photograph that measures 7 inches high by 10 inches wide is placed in a frame. Each side of the frame is 1 inch wide.

7 in.

10 in.

   a. What are the outer dimensions of the frame?

      height:_____ width:_____

   b. Is the rectangle formed by the outer edge of the frame similar to the rectangular photograph?

4. David wants to reduce a photograph to wallet size. At most, the smaller copy can be 4 inches high. At this height, what will be the width of the wallet-size copy?

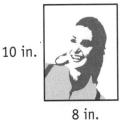

10 in.

8 in.
Original Photograph

4 in.

?
Wallet-Size Copy

5. On a United States map having a scale that reads "1 inch = 300 miles," Kelly measures the straight-line distance from Miami to Boston as $4\frac{1}{8}$ inches. What is the approximate distance between these two cities?

6. New York and Chicago are about 660 miles apart. How many inches (to the nearest quarter inch) will these cities be apart on a map on which the scale reads "1 inch = 200 miles"?
   (**Hint:** To find map distance, divide the actual distance by the map scale ratio.)

Surface Area

A three-dimensional object such as a box (rectangular solid) has both volume and surface area. The **surface area** of a box is the sum of the areas of the six outer surfaces. The box pictured below is 12 inches long, 3 inches wide, and 6 inches high. The box has six surfaces. These surfaces are best seen if the box is drawn unfolded.

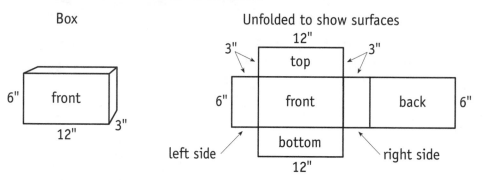

Box   Unfolded to show surfaces

The surface area is the sum of the areas of the six surfaces.

Surface	Area (sq in.)
left side:	$6 \times 3 = 18$
right side:	$6 \times 3 = 18$
front:	$12 \times 6 = 72$
back:	$12 \times 6 = 72$
top:	$12 \times 3 = 36$
bottom:	$12 \times 3 = 36$

> **Math Tip**
> The formula for the surface area of a cube is
> surface area = $6s^2$
> Can you explain why?

ANSWER: Total surface area = 252 sq in.

A box measures 6 feet long, 4 feet wide, and 3 feet high.

1. **a.** Label each surface with its dimensions.

3 ft   front   4 ft
6 ft

left side	right side	front	back	top	bottom

b. What is the surface area of the box?

2. **a.** What is the volume of the block shown here?

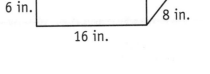

6 in. 8 in. 16 in.

   b. What is the surface area of the block?

3. **a.** What is the volume of the suitcase pictured at the right? Express your answer in cubic feet. (**Hint:** 8 in. = $\frac{2}{3}$ ft)

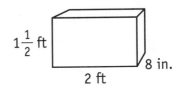

$1\frac{1}{2}$ ft 8 in. 2 ft

   b. What is the surface area of the suitcase? Express your answer in square feet.

4. The pyramid at the right has a square base and 4 triangular sides. Counting the base, what is the surface area of this pyramid?

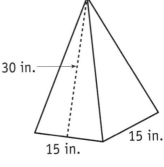

30 in. 15 in. 15 in. 15 in.

5. A cylindrical tank is 5 feet long. The *diameter* of the tank is 3 feet. To the nearest square foot, what is the surface area of the tank? (**Hint:** The length of the rectangle is the circumference of the circle.)

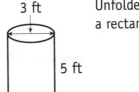

3 ft 5 ft

Unfolded, a cylinder can be pictured as a rectangle and 2 circles.

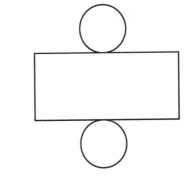

Using Algebra in Geometry

....

Algebra is often used to solve problems in geometry. In the examples below, unknown angles are represented by variables, and equations are used to find the measure of each angle.

From geometry you know that an angle ($\angle$) is measured in units called **degrees** (°). A right angle ($\llcorner$), formed by perpendicular lines, contains 90°. An angle can be labeled with three letters, the second letter being the **vertex,** the point where the two sides meet.

EXAMPLE 1 $\angle$ABC is a right angle (90°) that is divided into three smaller angles:

$\angle$DBE = 2 times $\angle$EBC or 2$\angle$EBC and
$\angle$ABD = 3 times $\angle$EBC or 3$\angle$EBC.

What is the measure in degrees of each of these smaller angles?

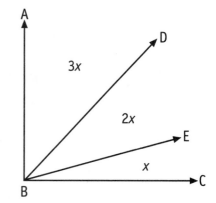

STEP 1 Let $\angle$EBC = x, $\angle$DBE = 2x, and $\angle$ABD = 3x.

STEP 2 Set the sum of the angles equal to 90° and solve the equation.

$$x + 2x + 3x = 90$$
$$6x = 90$$
$$x = \frac{90}{6} = \mathbf{15}$$
$$2x = 2(15) = \mathbf{30}$$
$$3x = 3(15) = \mathbf{45}$$

ANSWER: $\angle$EBC = 15°, $\angle$DBE = 30°, and $\angle$ABD = 45°

For Example 2, you need to remember that the sum of the three angles in a triangle always equals 180°. Also, an angle in a triangle is often represented by one letter only.

EXAMPLE 2 In $\triangle$ABC, $\angle$B = 3$\angle$A, and $\angle$C = 5$\angle$A. How many degrees is each angle?

STEP 1 Let $\angle$A = x, $\angle$B = 3x, and $\angle$C = 5x.

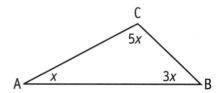

STEP 2 Set the sum of the angles equal to 180°, and solve the equation.

$$x + 3x + 5x = 180$$
$$9x = 190$$
$$x = \frac{180}{9} = \mathbf{20}$$
$$3x = 3(20) = \mathbf{60}$$
$$5x = 5(20) = \mathbf{100}$$

ANSWER: $\angle$A = 20°, $\angle$B = 60°, and $\angle$C = 100°

In each problem, solve for each angle indicated.

1. $\angle ABC = 90°$
   $\angle DBE = 3\angle EBC$
   $\angle ABD = 5\angle EBC$

   What is the measure in degrees of each angle?

   $\angle EBC =$

   $\angle DBE =$

   $\angle ABD =$

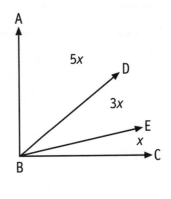

2. $\angle D + \angle E + \angle F = 180°$
   $\angle E = 3\angle D$
   $\angle F = 2\angle D$

   What is the measure in degrees of each angle?

   $\angle D =$

   $\angle E =$

   $\angle F =$

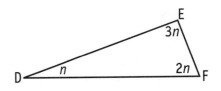

3. $\angle ABC = 90°$
   $\angle DBE = 5\angle EBC$
   $\angle ABD = 9\angle EBC$

   What is the measure in degrees of each angle?

   $\angle EBC =$

   $\angle DBE =$

   $\angle ABD =$

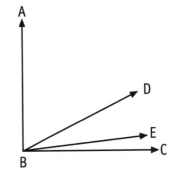

4. $\angle A + \angle B + \angle C = 180°$
   $\angle B = 4\angle A$
   $\angle C = 5\angle A$

   What is the measure in degrees of each angle?

   $\angle A =$

   $\angle B =$

   $\angle C =$

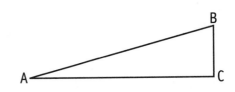

More About Exponents

Powers: Positive and Negative Exponents

As you've seen, you can use the exponent 2 to write the square of a number.

You can also use larger **positive exponents** as a shorthand way to indicate more **factors** (numbers being multiplied). The exponent tells how many times to write the base as a factor when you multiply.

Examples	As a base and an exponent	Read in words	Value
$5 \times 5 \times 5$	5^3	"5 to the 3rd power"*	125
$6 \times 6 \times 6 \times 6$	6^4	"6 to the 4th power"	1,296
$10 \times 10 \times 10 \times 10 \times 10$	10^5	"10 to the 5th power"	100,000

*5^3 is also called "5 cubed."

A **negative exponent** represents the number found by *inverting* the number represented by a positive exponent. Compare the examples below with the examples above.

Examples	As a base and an exponent	Read in words	Value
$\frac{1}{5 \times 5 \times 5}$	5^{-3} or $\frac{1}{5^3}$	"5 to the minus 3rd power"	$\frac{1}{125}$
$\frac{1}{6 \times 6 \times 6 \times 6}$	6^{-4} or $\frac{1}{6^4}$	"6 to the minus 4th power"	$\frac{1}{1,296}$
$\frac{1}{10 \times 10 \times 10 \times 10 \times 10}$	10^{-5} or $\frac{1}{10^5}$	"10 to the minus 5th power"	$\frac{1}{100,000}$ or 0.00001

Complete the following table.

	As a base and an exponent	Read in words	Value
1. $10 \times 10 \times 10 \times 10$	_____	_____	_____
2. $\frac{1}{3 \times 3 \times 3}$	_____	_____	_____
3. $\frac{1}{10 \times 10 \times 10 \times 10}$	_____	_____	_____

Find each value.

4. $7^3 =$ $2^{-5} =$ $10^6 =$ $10^{-3} =$

Scientific Notation

Scientific notation is a way to write numbers that contain many zeros. In scientific notation, a number is written as the product of two factors: a number between 1 and 10 *and* a power of 10.

The power tells how many places you move the decimal point to write the number using digits.

- A positive exponent tells you to move the decimal point to the right.
- A negative exponent tells you to move the decimal point to the left.

Positive Exponents

Negative Exponents

EXAMPLE 1 $8 \times 10^4 = 80,000$

four decimal places to the right

EXAMPLE 2 $3 \times 10^{-4} = 0.0003$

four decimal places to the left

Number	In Scientific Notation	Number	In Scientific Notation
90,000,000	9×10^7	0.00006	6×10^{-5}
570,000	5.7×10^5	0.00085	8.5×10^{-4}
753,000,000	7.53×10^8	0.000096	9.6×10^{-5}

....

Using a positive exponent, write each number in scientific notation.

5. $4,000 =$ $9,000,000,000 =$ $75,000,000 =$

6. $8,000,000 =$ $25,000 =$ $1,380,000 =$

Using a negative exponent, write each number in scientific notation.

7. $0.006 =$ $0.000008 =$ $0.0000925 =$

8. $0.0002 =$ $0.00075 =$ $0.00000318 =$

Solve each problem below.

9. The distance from Earth to the moon is approximately 240,000 miles. Write this distance in scientific notation.

10. Using a microscope, a biologist measured the diameter of a pollen grain to be 0.038 centimeter. Write this diameter in scientific notation.

11. The distance from Earth to the sun is about 9.3×10^7 miles. Write this distance as a whole number.

12. The diameter of a hydrogen atom is about 2×10^{-8} centimeter. Write this diameter as a decimal.

Using a Calculator

A calculator is a valuable math tool. You'll use it mainly to add, subtract, multiply, and divide quickly and accurately. The calculator pictured at the right is similar to one you've seen or may be using.

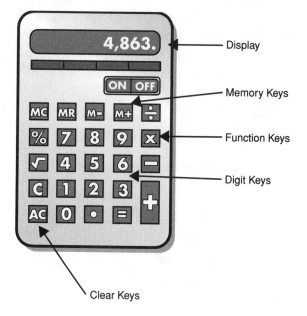

To enter a number, press one digit key at a time. On the display, the number 4,863 is entered. Notice the following features:

- A decimal point is displayed to the right of the ones digit.

- The calculator does not have a comma key or a dollar sign key.

- Pressing the clear key [C] erases the display. You should press [C] each time you begin a new problem or when you've made a mistake.

EXAMPLE 1   Divide 408 by 12.

STEP 1   Press [C] to clear the display.

STEP 2   Press the digit keys, divide key [÷], and equals key [=].

Press Keys   **Display Shows**

[C] [4] [0] [8] [÷] [1] [2] [=]   34.

ANSWER: 34

EXAMPLE 2   Multiply $12.75 by 6.

STEP 1   Press [C] to clear the display.

STEP 2   Press the digit keys, decimal point key, times key [×], and equals key [=].

Press Keys   **Display Shows**

[C] [1] [2] [·] [7] [5] [×] [6] [=]   76.5

ANSWER: $76.50

Press the decimal point key to separate dollars from cents.

Calculators do not show zeros at the right end of a decimal fraction.

Using Mental Math

Multiplying by Powers of Ten

The numbers 10, 100, 1,000, 10,000, and so on are called **powers of ten.** (They are so-named because $10 = 10^1$, $100 = 10^2$, $1,000 = 10^3$, $10,000 = 10^4$, and so on.) To multiply any number by a power of ten, count the zeros in the power of ten and write that number of zeros to the right of the other factor.

EXAMPLES

$$\begin{array}{r} 21 \\ \times\,1,000 \\ \hline 21,000 \end{array}$$ ← three zeros
← product

$$\begin{array}{r} 1,525 \\ \times\quad 100 \\ \hline 152,500 \end{array}$$ ← two zeros
← product

To multiply a decimal number by a power of ten, count the zeros in the power of ten. In the other factor, move the decimal point to the right that number of places.

EXAMPLES $13.52 \times 1,000 = 13,520$

three zeros   three places

$0.05238 \times 100 = 5.238$

two places   two zeros

To multiply two factors with zeros at the right, count the zeros in both factors. Ignoring all the zeros, multiply the two numbers. Then write that number of zeros to the right of the product.

EXAMPLES

$$\begin{array}{r} 130 \\ \times\quad 200 \\ \hline 26,000 \end{array}$$ ← three zeros

13×2 three zeros

$$\begin{array}{r} 1500 \\ \times\quad 4000 \\ \hline 6,000,000 \end{array}$$ ← five zeros

15×4 five zeros

Dividing by Powers of Ten

To divide by a power of ten, count the zeros in the power of ten. In the numerator (the dividend), move the decimal point to the left that number of places.

EXAMPLES $\dfrac{575.2}{100} = 5.752$

two zeros   two places

$\dfrac{1.32}{10,000} = 0.000132$

four zeros   four places

To divide two numbers with zeros at the right, cancel the same number of zeros in the numerator as in the denominator.

EXAMPLES $\dfrac{600}{30} = \dfrac{60\cancel{0}}{3\cancel{0}} = \dfrac{60}{30} = 20$ $\dfrac{7,000}{400} = \dfrac{700\cancel{0}}{40\cancel{0}} = \dfrac{70}{4} = 17.5$

Using Estimation

Front-End Estimation

In **front-end estimation,** you round each number to its left-most digit. Then you can use mental math to calculate with the rounded values.

<u>ADDITION EXAMPLE</u>   Find the rounded values by rounding each number to its left-most digit. Then add the rounded values.

Exact Problem	**Rounded Values**
687	700
4,435	4,000
+ 393	+ 400
	5,100

ANSWER: The estimate is 5,100.

<u>SUBTRACTION EXAMPLE</u>

Exact Problem	**Rounded Values**
483,478	500,000
− 217,675	− 200,000
	300,000

ANSWER: The estimate is 300,000.

<u>MULTIPLICATION EXAMPLE</u>

Exact Problem	**Rounded Values**
385	400
× 57	× 60
	24,000

ANSWER: The estimate is 24,000.

<u>DIVISION EXAMPLE</u>   Instead of rounding to the left-most digit, replace the numbers with approximate values that make the new problem divide evenly.

Exact Problem	**Approximate Values**	**Reduced**	**Estimate**
$\dfrac{529,420}{710}$	$\dfrac{490,0\cancel{0}\cancel{0}}{7\cancel{0}\cancel{0}}$	$\dfrac{4900}{7}$	700

ANSWER: The estimate is 700.

Formulas

PERIMETER

Figure	Name	Formula	Meaning
	Rectangle	$P = 2l + 2w$	l = length w = width
	Square	$P = 4s$	s = side
	Triangle	$P = s_1 + s_2 + s_3$	s_1 = side 1 s_2 = side 2 s_3 = side 3
	Polygon (n sides)	$P = s_1 + s_2 + \ldots s_n$	s_1 = side 1, side 2, and so on
	Circle (circumference)	$C = \pi d$ or $C = 2\pi r$	$\pi \approx 3.14$ or $\frac{22}{7}$ d = diameter r = radius

AREA

Figure	Name	Formula	Meaning
	Rectangle	$A = lw$	l = length w = width
	Square	$A = s^2$	s = side
	Parallelogram	$A = bh$	b = base h = height
	Triangle	$A = \frac{1}{2}bh$	b = base h = height
	Circle	$A = \pi r^2$	$\pi \approx 3.14$ or $\frac{22}{7}$ r = radius

VOLUME

Figure	Name	Formula	Meaning
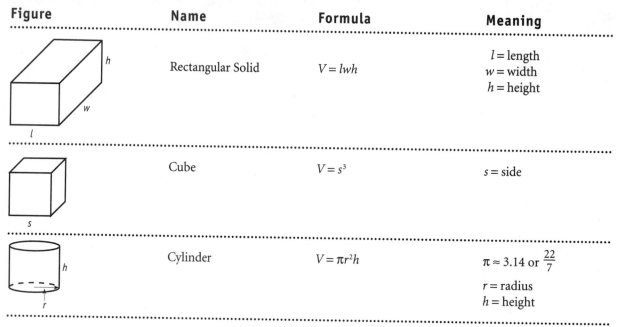	Rectangular Solid	$V = lwh$	l = length w = width h = height
	Cube	$V = s^3$	s = side
	Cylinder	$V = \pi r^2 h$	$\pi \approx 3.14$ or $\frac{22}{7}$ r = radius h = height

ANGLES

Name	Figure	Formula	Meaning
Sum of angles in a triangle		$\angle a + \angle b + \angle c = 180°$	a, b, and c are the three angles of any triangle
Pythagorean theorem		$c^2 = a^2 + b^2$	c = hypotenuse a and b are the two shorter sides of a right triangle

OTHER FORMULAS

Name	Formula	Meaning
Average (Mean)	$\dfrac{a + b + c + \ldots}{n}$	a, b, c, . . . are individual values in a set of values. n = the number of numbers in the set
Percent	$P = W \times \%$ $W = \dfrac{P}{\%}$ $\% = \dfrac{P}{W}$	Part = Whole × Percent Whole = Part ÷ Percent Percent = Part ÷ Whole

Glossary

A

acute angle   An angle measuring more than 0° but less than 90°

acute angle

algebraic expression   Two or more numbers or variables combined by addition, subtraction, multiplication, or division

angle   Figure formed when two rays have the same endpoint, called the vertex of the angle

angles

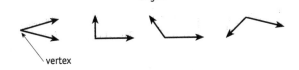

vertex

area (A)   An amount of surface

3 yd

4 yd

$$\text{area} = l \times w$$
$$= 4 \times 3$$
$$= 12 \text{ square yards}$$

area unit   Square units that are used to measure area. Common area units are square inches, square feet, square yards, and square meters.

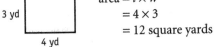

area unit

average   A middle value of a set of values. The arithmetic average (mean) is found by adding the set and then dividing the sum by the number of values in the set.

Stacey received three math scores: 83, 78, and 91. Stacey's average score is 84.

$$83 + 78 + 91 = 252$$
$$252 \div 3 = 84$$

axes   The perpendicular sides of a graph along which numbers, data values, or labels are written

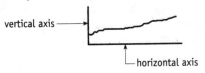

vertical axis

horizontal axis

B

bar graph   A graph that uses bars to display information. Data bars can be drawn vertically or horizontally.

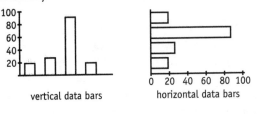

vertical data bars   horizontal data bars

base   The number being multiplied in a power. For example, in 4^3, 4 is the base.

binomial   A polynomial of two terms. For example, $2x + 6$.

C

capacity   The volume of liquid (or substance such as sugar) that a container can hold. Common capacity units are fluid ounce, quart, gallon, milliliter, and liter.

centimeter   A metric unit of length equal to 0.01 meter. 1 inch = 2.54 centimeters

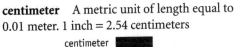

centimeter

inch

actual size

circle   A plane (flat) figure, each point of which is an equal distance from the center

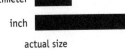

center

circle

circle graph   A graph that uses a divided circle to show data. Each part of the circle is called a *sector* or a *section*. The sections add up to a whole or to 100%. A circle graph is also called a *pie graph* or *pie chart*.

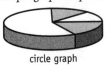

circle graph

circumference The distance around a circle
Circumference = π × diameter (π ≈ 3.14)

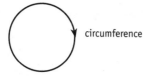

circumference

column In a table, a vertical listing of numbers or words that is read from top to bottom

column of numbers	column of words
109	Monday
175	Tuesday
206	Wednesday
214	Thursday

comparison symbols Symbols used to compare one number with another. Examples: < and >.

Symbol	Meaning	Example	
<	is less than	$5 < 8$	5 is less than 8.
>	is greater than	$7 > -3$	7 is greater than −3.
≤	is less than or equal to	$n \leq 5$	n is less than or equal to 5.
≥	is greater than or equal to	$x \geq 3$	x is greater than or equal to 3.

complementary angles Angles whose sum is 90°

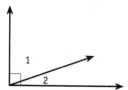

∠1 and ∠2 are complementary angles.

composite number A number that has more than two factors

congruent Having exactly the same size and shape

congruent figures

coordinate axis A number line used as either the horizontal or vertical axis of a graph. See **horizontal axis** and **vertical axis.**

corresponding angles Pairs of equal angles in similar or congruent figures.

Similar Triangles

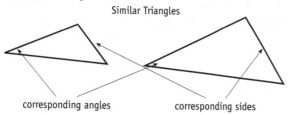

corresponding angles   corresponding sides

corresponding sides In similar figures, sides that are opposite corresponding angles. See **corresponding angles.**

cross products In a proportion, the product of the numerator of one ratio times the denominator of the other. In a true proportion, cross products are equal.

Proportion	Equal Cross Products
$\frac{2}{3} = \frac{6}{9}$	$2 \times 9 = 3 \times 6$
	18 = 18

cube A 3-dimensional shape that contains six square faces. At each vertex, all sides meet at right angles.

face

cube

cylinder A 3-dimensional shape that has both a circular base and a circular top

cylinder

D

data A group of numbers or words that are related in some way

Numbers: $2.50, $3.75, $6.40
Words: beef, chicken, fish, pork

degrees The measure (size) of an angle. A circle contains 360 degrees (360°). One-fourth of a circle contains 90°.

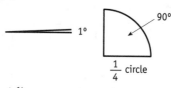

1°   90°

$\frac{1}{4}$ circle

diagonal A line segment running between two non-consecutive vertices of a polygon. A diagonal divides a square or rectangle into two equal triangles.

diagonal

diameter A line segment, passing through the center, from one side of a circle to the other. The length of the diameter is the distance across the circle.

diameter

distance formula The formula $d = rt$ that relates distance, rate, and time. In words, distance equals rate times time. By rearranging the variables, you can write the rate formula $r = \frac{d}{t}$ or the time formula $t = \frac{r}{d}$.

double bar graph   A bar graph containing two sets of bars to display and compare two sets of related data

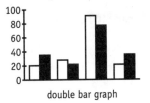

double bar graph

double line graph   A line graph containing two lines to display and compare two sets of related data

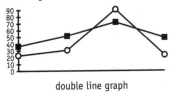

double line graph

E

equation   A statement that says two amounts are equal (have equal value or measure)

equilateral triangle   A triangle with three equal sides and three equal angles, each measuring 60°. An equilateral triangle is a regular polygon.

equilateral triangle

exponent   A number that tells how many times the base (of a power) is used as a factor. For example, in 5^2, 2 is the exponent.

extrapolate   To make a reasonable guess of a data value that lies outside a given set of values

Four given data values: 3, 6, 9, 12
Extrapolated fifth value: 15

F

factor   A number that divides evenly into another number. Example: 1, 2, 4, and 8 are factors of 8.

fluid ounce   A small unit of capacity. One cup is 8 fluid ounces; 1 quart is 32 fluid ounces.

foot   A unit of length equal to 12 inches. There are 3 feet in 1 yard.

G

gallon   A unit of capacity equal to 4 quarts. One gallon is 128 fluid ounces.

H

horizontal   Running right and left

horizontal axis   On a graph, the axis running left to right

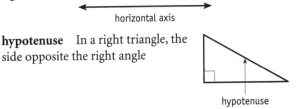

horizontal axis

hypotenuse   In a right triangle, the side opposite the right angle

hypotenuse

I

inequality   A statement that compares two numbers

intercept   A number that tells where a graphed line crosses a coordinate axis

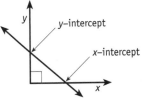

interpolate   To estimate or guess a data that lies between two known values

isosceles triangle   A triangle in which two sides have the same length. The two angles opposite the equal sides have the same measure. The two equal angles are called *base angles*.

base angles

isosceles triangle

K

kilometer   A metric unit of length equal to 1,000 meters. 1 meters ≈ 0.6 miles

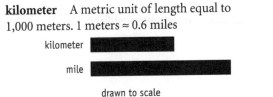

drawn to scale

L

line   A path of points that extends in two opposite directions. A line has no endpoints.

line

linear equation   An equation whose graph is a single straight line. Example: $y = 2x + 3$

line graph A graph that displays data as points along a graphed line

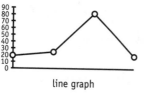

line graph

line segment Part of a line having two endpoints

line segment

M

mean The sum of the data values, divided by the number of data values. For the data values 1, 3, 3, 4, the mean is $\frac{1+3+3+4}{4} = \frac{11}{4} = 2.75$.

median The middle value in an ordered set of data values

Example: 2 7 9 9 11 15 15
↑ median

Example: 10, 12, 16, 16
$(12 + 16) \div 2 = 14$

meter A metric unit of length equal to 100 cm. 1 meter ≈ 39.6 inches, a little longer than 1 yard

meter

yard

drawn to scale

metric system The measurement system whose basic units are meter (for length), liter (for capacity), and gram (for weight)

mode The most common or most frequent data value

Example: 5, 6, 6, 7, 7, 7, 8, 8   The mode is 7.

Example: 4, 4, 5, 6, 6, 7, 8 10   The mode is 4 or 6.

N

negative numbers Numbers less than zero. A negative number is written with a negative sign.
Examples: −6, −2.5, −1

number line A line used to represent both positive and negative numbers

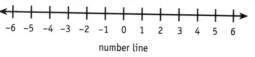

number line

O

obtuse angle An angle that measures more than 90° but less than 180°

obtuse angle

ordered pair A pair of numbers written within parentheses, such as (2, 3)

P

parallel lines Lines in the same plane that do not meet

parallel lines

parallelogram A four-sided polygon with two pairs of parallel sides. Opposite sides are equal and opposite angles have equal measure.

parallelogram

percent Part of 100. For example, 5 percent means 5 parts out of 100.

perfect square A number whose square root is a whole number. Example: 49 is a perfect square because $\sqrt{49} = 7$.

perimeter The distance around a flat (plane) figure. The symbol for perimeter is P.

perpendicular lines Lines that meet (or cross) at a right angle (90°)

perpendicular lines

pi (π) The ratio of the circumference of a circle to its diameter. Pi is approximately 3.14 or $\frac{22}{7}$.

pictograph A graph that uses small pictures or symbols to represent data. Data lines may be displayed either horizontally or vertically.

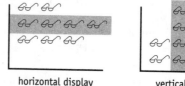

horizontal display   vertical display

pie graph See **circle graph.** (Pie graph is another name for circle graph.)

plane figure A 2-dimensional (flat) figure. Examples: circles, rectangles, and triangles

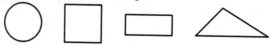

polygon A plane (flat) figure formed by line segments that meet only at their endpoints

polynomial One or more algebraic terms combined by addition or subtraction

power The product of a number multiplied by itself one or more times. Example: $3^2 = 3 \times 3$

prediction Regarding a graph, an estimate about the value of an unknown data point

proportion Two equal ratios. A proportion can be written with colons or as equal fractions.

Written with colons: $2:3 = 6:9$

Written as equal fractions: $\frac{2}{3} = \frac{6}{9}$

Pythagorean theorem A theorem that states: In a right triangle, the square of the hypotenuse is equal to the sum of the squares of the two remaining sides.

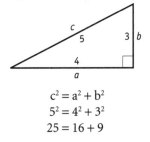

$$c^2 = a^2 + b^2$$
$$5^2 = 4^2 + 3^2$$
$$25 = 16 + 9$$

R

radius A line segment from the center of a circle to any point on the circumference of the circle

radius

ratio A comparison of two numbers. A ratio can be written as a fraction or with a colon.

5 to 4 is written as $\frac{5}{4}$ or 5:4

ray Half of a line. A ray has one endpoint. An angle is formed when two rays have the same endpoint.

endpoint

ray

rectangle A four-sided polygon with two pairs of parallel sides and four right angles. The opposite sides of a rectangle have equal length.

rectangle

rectangular solid (prism) A 3-dimensional figure in which each face is either a rectangle or a square. Opposite faces are congruent.

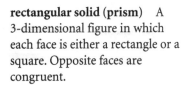

reflex angle An angle that measures more than 180° but less than 360°

reflex angle

right angle An angle that measures exactly 90°. A right angle is often called a square corner.

right angle

right triangle A triangle that contains a right angle

right triangle

row In a table, a horizontal list of numbers or words that is read from left to right

numbers: 14, 19, 26, 31

words: pennies, nickels, dimes, quarters

S

scalene triangle A triangle in which no two sides are congruent and no two angles are congruent

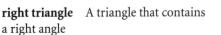

scalene triangle

signed numbers Negative numbers and positive numbers

similar triangles Triangles that have three equal angles. Similar triangles have the same shape and may differ only in the lengths of their sides.

similar triangles

slope The amount by which a straight line rises or falls in one unit of horizontal distance

sphere A 3-dimensional shape, each point of which is an equal distance from the center. A basketball is a sphere.

sphere

square   A regular polygon with four equal sides, two pairs of parallel sides, and four right angles

square

square of a number   The product of a number multiplied by itself. Example: $8^2 = 8 \times 8 = 64$

square root ($\sqrt{\ }$)   One of two equal factors of a number. Example: $\sqrt{9} = 3$

straight angle   An angle that measures exactly 180°, having the shape of a straight line

straight angle

supplementary angles   Angles that add to 180°

∠3 and ∠4 are supplementary angles.

T

three-dimensional figures   Figures that take up space and have volume. Examples are cubes, rectangular solids, and cylinders.

transversal   A line that cuts across two or more other lines, intersecting each of them

transversal

triangle   A polygon with three sides and three angles

triangle

two-dimensional figure   See **plane figure.**

V

vertex   The point where the two sides of an angle meet

vertex

vertical angles   Angles that are formed when two lines cross. Vertical angles lie across from each other and are equal.

∠5 and ∠7 are vertical angles.
∠6 and ∠8 are vertical angles.

vertical axis   On a graph, the axis running up and down

vertical axis

volume   The amount of space an object takes up. Volume is usually measured in cubic units.

volume unit   Cubic units that are used to measure volume. Common volume units are cubic inches, cubic feet, cubic yards, cubic centimeters, and cubic meters.

volume unit

Index